Ordinal Data Analysis

This book is a step-by-step data story for analyzing ordinal data from start to finish. The book is for researchers, statisticians and scientists who are working with datasets where the response is ordinal. This type of data is common in many disciplines, not just in surveys (as is often thought). For example, in the biological sciences, there is an interest in understanding and predicting the (growth) stage (of a plant or animal) based on a multitude of factors. Likewise, ordinal data is common in environmental sciences (for example, stage of a storm), chemical sciences (for example, type of reaction), physical sciences (for example, stage of damage when force is applied), medical sciences (for example, degree of pain) and social sciences (for example, demographic factors like social status categorized in brackets). There has been no complete text about how to model an ordinal response as a function of multiple numerical and categorical predictors. There has always been a reluctance and reticence towards ordinal data as it lies in a no-man's land between numerical and categorical data.

Examples from health sciences are used to illustrate in detail the process of how to analyze ordinal data, from exploratory analysis to modeling, to inference and diagnostics. This book also shows how Likert-type analysis is often used incorrectly and discusses the reason behind it. Similarly, it discusses the methods related to Structural Equations and talks about appropriate uses of this class of methods.

The text is meant to serve as a reference book and to be a "how-to" resource along with the "why" and "when" for modeling ordinal data.

Key Features:

- Includes applications of the statistical theory
- Includes illustrated examples with the associated R and SAS code
- Discusses the key differences between the different methods that are used for ordinal data analysis
- Bridges the gap between methods for ordinal data analysis used in different disciplines

Nairanjana (aka "Jan") Dasgupta is a Regents Professor and Boeing Distinguished Professor of Math and Science Education at Washington State University. She also serves as the Professor of Statistics and Director of the Data Analytics Program. She is a Fellow of ASA and an inductee in the WA State Academy of Sciences. She is passionate about the understandability and applicability of statistical methods in real-life applications.

Jillian Morrison is Assistant Professor of Statistical and Data Sciences at The College of Wooster. She serves as Chair of the Communications Committee and a member of the governing council for the Caucus for Women in Statistics and Data Science. She is also a fellow of the Mathematical Association of America's Project NExT. Her work focuses on applying statistical and data science methods to solve problems in the real world.

Ordinal Data Analysis

Statistical Perspective with Applications

Nairanjana Dasgupta
Jillian Morrison

CRC Press is an imprint of the
Taylor & Francis Group, an **informa** business
A CHAPMAN & HALL BOOK

Designed cover image: © Shutterstock Stock Photo ID 1924361840, Photo Contributor Dilok Klaisataporn

First edition published 2024
by CRC Press
2385 NW Executive Center Drive, Suite 320, Boca Raton FL 33431

and by CRC Press
4 Park Square, Milton Park, Abingdon, Oxon, OX14 4RN

CRC Press is an imprint of Taylor & Francis Group, LLC

ISBN: 978-0-367-85590-1 (hbk)
ISBN: 978-1-032-76423-8 (pbk)
ISBN: 978-1-003-02061-5 (ebk)

DOI: 10.1201/9781003020615

Typeset in CMR10
by KnowledgeWorks Global Ltd.

Publisher's note: This book has been prepared from camera-ready copy provided by the authors.

Baba and Ma
For insisting I write this
And Dean, Meghna and Tage
For always being there. ND

To my parents,
Who planted the seed of knowledge in me
And nurtured it.
To Moses and Karis,
For supporting me always. JM

Contents

Preface xi

Acknowledgments xiii

I Introduction to Ordinal Data 1

1 Ordinal Data 3
- 1.1 Introduction 3
- 1.2 Motivating Datasets 5
  - 1.2.1 Horse Data: EIPH in Race Horses 5
  - 1.2.2 Heart Data: Coronary Atherosclerosis 6
- 1.3 Questions Arising in Ordinal Data 6
- 1.4 Software Used: R and SAS 7

II Exploratory Analysis 9

2 Summarizing and Visualizing Ordinal Data 11
- 2.1 Introduction 11
- 2.2 Graphical Summary 11
  - 2.2.1 Univariate Ordinal Data 12
  - 2.2.2 Two Variables: Ordinal Response and Categorical Explanatory Variable 14
  - 2.2.3 Two Variables: Ordinal Response and Ordinal Explanatory Variable 17
  - 2.2.4 Two Variables: Ordinal Response and Discrete Numerical Explanatory Variable 19
  - 2.2.5 Two Variables: Ordinal Response and Continuous Explanatory Variable 20
- 2.3 Numerical Summary 22
  - 2.3.1 Univariate Ordinal Data 23
  - 2.3.2 Two Variables: Ordinal Response and Categorical Explanatory Variable 24
  - 2.3.3 Two Variables: Ordinal Response and Ordinal Explanatory Variable 26
  - 2.3.4 Two Variables: Ordinal Response and Discrete Numerical Explanatory Variable 28

2.3.5 Two Variables: Ordinal Response and Continuous Explanatory Variable 29
2.4 R Syntax 30
2.5 SAS Syntax 32
2.6 Summary 33

III Methods for the Analysis of Ordinal Data 35

3 Historical Perspective 37
3.1 Likert Method 38
3.2 Cumulative Ordinal Models 40
3.3 Latent Variable Models (LVM) 41
3.4 Rank-Based Methods 43
3.5 Applications of the Different Methods 43
3.6 Summary 45

4 Likert Scale 46
4.1 Introduction 46
4.2 Likert Items versus Likert Scale 46
4.3 Data Analysis 48
4.4 Summary 51

5 Cumulative Distribution Function (CDF) Models 52
5.1 Introduction 52
5.2 Models 53
5.2.1 General Linear Model 53
5.2.2 Generalized Linear Model 53
5.2.3 The CDF Models 53
5.3 Theory of CDF Models 54
5.3.1 CDF as Threshold Models 54
5.3.2 Logistic Distribution 55
5.3.3 Ordinal Logistic Regression 57
5.3.4 Understanding the Parameters Mathematically 59
5.3.5 Proportional Odds Assumption 60
5.3.6 Estimating the Parameters: Maximum Likelihood Estimation 62
5.3.7 Inference: Confidence Intervals and Testing 64
5.3.8 Types of Logits 65
5.3.9 Ordinal Probit Regression 66
5.4 Data Analysis 67
5.5 Summary 85

6 Latent Variable Models: Structural Equation Models 86
6.1 Introduction 86
6.2 Structural Equation Modeling 88
6.3 The Models 89
6.3.1 Simple Linear Regression: Relating One X to One Y 89
6.3.2 Multiple Linear Regression: Relating Multiple X's to One Y 90
6.3.3 Multivariate Regression: Relating Multiple X's to Multiple Y's 91
6.3.4 Path Analysis: Endogenous to Endogenous, Relating Response Y's to Each Other 92
6.3.5 Confirmatory Factor Analysis: Relating Latent Response(s) with Observed **x** 93
6.3.6 Structural Equations: Latent to Latent Relationship . 96
6.4 Structural Equation Modeling for Ordinal Data 99
6.5 Summary 101

IV Further Work with Ordinal Data 103

7 Diagnostics of the Ordinal Regression Models 105
7.1 Introduction 105
7.2 Assumptions for Ordinal Logistic Model 106
7.3 Goodness of Fit Tests 108
7.3.1 Hosmer and Lemeshow Test 108
7.3.2 The Pulkstenis-Robinson (P-R) Test 109
7.3.3 The Lipsitz Test 110
7.4 Association Statistics 112
7.4.1 Percent Concordant and Percent Discordant 112
7.4.2 Somers' D 112
7.4.3 Goodman-Kruskal Gamma 113
7.4.4 Kendall's Tau-A 114
7.5 Model Fit Using Likelihood: AIC, BIC, and −2 Log L 115
7.6 Proportional Odds Assumption 117
7.6.1 Brant-Wald Test 117
7.6.2 Partial Proportional Odds Model 118
7.6.3 Other Models 120
7.7 Autocorrelation 120
7.7.1 Serial Dependence Measures 121
7.8 Multicollinearity 125
7.8.1 Variance Inflation Factor 125
7.8.2 Principal Component Regression 126
7.9 Summary 128

8 Simulating Ordinal Models **129**
8.1 Introduction 129
8.2 Simulating from a CDF Model 130
8.3 Ordinal Probit Structure 131
8.4 Ordinal Logistic Structure 139
8.5 Summary 145

V Overall Summary **147**

9 Data Story: Analysis of the Heart Data **149**
9.1 Introduction 149
9.2 Exploratory Data Analysis 149
9.3 Analysis Using Ordinal Logistic Model 155
9.4 Model Selection 158
9.5 Diagnostics 168
9.6 Summary 173

10 Tying up Loose Ends and Overall Summary **174**
10.1 Introduction 174
10.2 Findings from Datasets 175
10.3 Summary: Final Words 179

Bibliography **181**

Index **187**

Preface

The idea of this book came about after we were accosted by a slew of consulting problems related to ordinal regression. In each case, the premise was similar. We had an observational study with multiple explanatory variables and an ordinal response, which was not the result of a survey. As we started looking at the problems, we realized that no text focused on modeling and analysis that took one through the process of analyzing ordinal data from start to finish. We looked at the existing text on the topic from stalwart in the field, "Analysis of Ordinal Categorical Data" by Alan Agresti and realized that the focus of the book was more on designed experiments. Additionally, common questions from clients surrounded the applicability of using Likert's method or Structural Equation Modeling as was suggested to them. We thus began our journey of trying to get a handle on the problem from a statistical point of view. Our journey led to this book.

This book is for researchers, statisticians, and scientists who deal with ordinal regression. For example, this book could be used by social scientists trying to get a feel of how some of the methods used in social science fit in the framework of traditional statistics and vice versa. Because this book is a text for researchers in all areas, the the content is laid out without any slant coming from a specific field. One of our realizations after working with researchers is that people in the different fields are very siloed, and while researchers from social science and medical science might work on the same data related problem, there is not much exchange of information.

We have written this book from the context of two datasets (which were some of the problems that motivated our work) as a data story from start to finish with all the steps in between. In that story, we digress as we talk about methods like Likert and Path analysis (which were not strictly relevant for the datasets). We use R and R Studio and SAS studio software to show how to analyze the datasets using different analysis methods.

We hope that the book is readable and useful for all who in their data journey come across the problem of ordinal data.

Acknowledgments

We would like to thank David Grubbs and others from CRC Press for giving us the opportunity to write this book. It was a random conversation with David at the Joint Statistical Meetings that seeded the idea for this book. While COVID and other life changes came and went, David did not give up on us and the project. We are very grateful to him and his team.

We both are grateful to the Mathematics and Statistics Department at Washington State University, where JM was a PhD student under the supervision of ND which is where issues around working with ordinal data were discovered. We are grateful to Tory Schmidt and the WA Tree Fruit Research Commission for reintroducing us to the problem of ordinal data via problems related to apple stages. Our sincere gratitude to Dr. Warwick Bayley and his team for introducing the Horse Dataset and all its nuances. Author ND was part of the Spokane study for the Heart data. While the study was concluded over twenty years ago, we are grateful to Dr. Mielke and Dr. Broemling for sharing the study with her. Additional thanks to the Ministry of Education and high schools and parents in Belize for giving permission to collect data from high school students which is another dataset that allowed us to explore issues with ordinal data.

Various students of ND have contributed to the book. We would like to give special thanks to Dr. Swarnita Chakraborty, Dr. Adebo Sijuwade, and Menqi Yin for helping us with the write-up of LVM models, autocorrelation, LaTex, and R Code, and help with bibliography. We are grateful to Justice Nii-Ayitey, Dr. Debasmita Das, Lili Zhou, and Dr. Yingzhi Li for the feedback on methods. Bruce Austin deserves a special mention for the hours he spent helping us understand SEM models and pointing us to the seminal papers in the field.

ND: The professors at Washington State University deserve special mention. These include: Dr. Sam Saunders (late), Dr. Mike Jacroux (emiteritus), Dr. Rich Alldredge (emiteritus), Dr. Ron Mittellhammer, Dr. Krishna Jandhyala, and Dr. Marc Evans (emiteritus), who were mentors along the way. Dr. Jave Pascual, who was a peer, and Dr. Yuan Wang, Dr. Abhishek Kaul, and Dr. Xiongzhi Chen, my younger colleagues. Special mentions also go to Dr. Sandy Cooper and Dr. Lynn Schryer for the “girls lunches” that kept me sane.

Current faculty at the Department of Mathematics and Statistics all contributed with their patience and support.

I want to express my deep gratitude to my father Dr. Manas Dasgupta, a fellow academic, who insisted that I write a book on this topic. Along the way, he passed away in 2020, but his spirit kept me from giving up on the book. A special shout out to my late mother Dr. Amiya Dasgupta, who showed me what it meant to be a trail blazer as a woman. Deep unstinted gratitude to my late parents-in-law Mabelline and Harry Johnson for making me feel at home in a different country. Even when they did not understand my compulsion for research, they always were proud of me. I also give my thanks to my sisters Dr. Jhinuk Dasgupta and Dr. Sukti Dasgupta and my brother Mr. Neelanjan Dasgupta for always being in my corner and always being supportive and proud of me. My sisters-in-law Penny Grove and Carol Day and Dr. Betty Dasgupta for the sisterhood they provided and my brothers-in-law Buddy Day (late), Ken Grove, Joydeep Dasgupta, and Dr. Sriram Natarajan for being the protective older brothers always. If I list my extended family and friends, this book would exceed the page limit. It took a village, and I was grateful to all in that village.

My life would not be complete without my husband and soul mate Dr. H. Dean Johnson. He has always been the cheer-leader, the supporter, the confidant, the friend, and the partner in every aspect of my life. He has always facilitated every achievement in my life, with his love and support. My two children Meghna and Tage Johnson, who being here, allow me to keep my feet on the ground. Nothing would be possible without them loving me for who I was and all the hugs, kisses, sass, and attitude.

I am always grateful that Jillian (JM) picked Washington State University to come for her PhD and picked me as an advisor. Her energy, positivity, and can-do attitude make our collaborations fun. Using racing parlance "this book wouldn't have made it through the starting gate" without her. A special shout-out to Karis Luri (JM's daughter) who was born during this book project and brings us all so much joy.

JM: I express my deep gratitude to my parents, Douglas and Judy Morrison. Their endless support was instrumental in keeping me motivated to write when there were always a million other things to do. Special thanks to my brothers, Everett and Keron Morrison, who always made sure to remind me to keep writing.

To my husband, Moses Luri, thank you for your unwavering support and patience throughout this process. Also, to my daughter Karis Luri for being my little helper and making sure that I'm always checking for typos.

To Jan, thank you for always believing in me and supporting my endeavors. Your kindness and strength is unmatched.

Finally, thanks to my supporters and cheerleaders at the College of Wooster. The support you provided is greatly appreciated.

Part I

Introduction to Ordinal Data

1

Ordinal Data

1.1 Introduction

Traditional Statistics focuses on data being divided into two groups: categorical (qualitative) and numerical (quantitative). Further, numerical data is divided into two groups: discrete (countable) and continuous (measurable). Numerical continuous data (for example, height, weight, blood pressure, etc.) has received more attention than any other type in statistics. This is the corner-stone of classical statistics.

Discrete data (for example, the number of people living in a household or the number of genes expressed in cells by women with breast cancer, etc.) has received some attention recently (for example, texts by Santner and Duffy, 1989 [77], Friendly and Mayer, 2015 [27]). While there is less literature on discrete data than on continuous data, there **are** a few substantive books on the topic. The body of literature on categorical data (examples: profession, hair color), long neglected, is increasing in recent years. The publishing of substantive texts like the ones by Hosmer and Lemeshow (1989) [35] and Agresti (2010) [1] among others have helped increase popularity on this topic.

In this dichotomy (numerical or categorical) there really is no defined place for ordinal data. It is assumed to be categorical by some and discrete (numerical) by others. While this "eyes of the beholder" approach has been used for a long time, scientifically, it is not a reasonable approach. Discrete data is compiled, summarized, and analyzed very differently than categorical data. With categorical data being split among groups, measurements are usually done by counting the number of items that fall in a certain group or that have a certain characteristic, for example, the colors of M&M's. Notice that we cannot calculate the average of groups with categorical data. But with discrete data (for example, the number of cars at an intersection), we can calculate averages and standard deviations.

Another categorization of data (often used in social sciences) groups data as nominal, ordinal, interval, and ratio. This specification includes ordinal data. However, in this specification too, it is unclear how to analyze and predict using ordinal data. This is somewhat interesting as ordinal data has been

DOI: 10.1201/9781003020615-1

collected from time immemorial. As long as people have ordered their responses (for example, degree of pain, stages of growth, etc.), we have had ordinal data. Even before the days of the Likert method (1932) [51], people have been thinking about this topic. However, attention on this topic in statistics has waxed and waned. One reason for this lack of interest is that people in different fields have dealt with ordinal data differently, and it has been accepted in their field. As such they see no reason to change their approach. Though there has always been controversy and divide about how to analyze and use ordinal data, it has not garnered as much attention in the statistics community. Our conjecture for this is that the different fields using ordinal data are often siloed away from each other. Hence, open discussion about the methods used has rarely happened. Our hope is a text devoted to ordinal data will fill this gap.

The interesting aspect of ordinal data is that it is what the name suggests: an ordering. While, the ordering is often done with discrete numbers, 1 through k, the numbers themselves do not mean anything except the order. The difference between the numbers are not relevant, nor is the ratio. For example, if our agreement scale is 1 to 5, it does not mean that the difference between 1 and 2 is the same as the difference between 3 and 4. The numbers are just place holders for an ordering. As the ordinal categories are often numbers, it is tempting to consider it discrete and use and analyze them as such, even though they are really categories. On the other hand, if considered as, we can take average and median of the ranks. Hence, unlike categorical data, this average rank can be calculated and it does provide some relevant information.

Ordinal data arises just as often in the laboratory sciences and health sciences. However, for historical reasons it is often associated with social sciences, especially with surveys. Surveys often use a scale of $1-K$. For example, on a scale of 1–5 (with 5 being the highest possible dislike), how do you feel about the current political climate? Other examples include online maps that collects data from its users about their ratings of places visited which is used to provide feedback to other users and business owners and used in their algorithms to make suggestions to future users. A common example in the health sciences is the pain-scale we are asked in hospital visits: rate our pain on a scale of 1–8. Developmental stage, tumor stage, and disease stage are all examples of ordinal outcomes. Health-related surveys are all too common. This type of data is also common in plant and animal sciences where stage of growth of plant or animal and order of outcomes in experiment are frequently used examples. In this book we will use two data examples (with the raw data provided from authors) to motivate and discuss the methods discussed.

The purpose of this book is modeling an ordinal response, when there are multiple types of explanatory variables collected. Hence, we stay away from datasets that have mostly categorical responses as seminal texts like Agresti

(2010) [1] talk about. Unlike "categorical" predictions which are classification problems, we consider ordinal predictions as a prediction problem. Essentially, the raison d'etre for this book is to understand this conundrum and look at ordinal data in the regression context.

Our purpose is to make this book very example oriented and hands-on. As a result, we use real datasets and provide the question-specific analysis of the dataset in various contexts. In the next section, we discuss our motivating datasets.

1.2 Motivating Datasets

Our first motivating dataset is an example about race horses with ordinal responses. Our second dataset looks at coronary disease risk. This book will follow these datasets through the data analysis process of exploratory data analysis, modeling, inference, and diagnostics.

1.2.1 Horse Data: EIPH in Race Horses

Exercise Induced Pulmonary Hemorrhage (EIPH) is a fairly common problem in race horses. This means that after a very strenuous exercise (like a race) horses tend to have some bleeding in the lungs. Roughly half to three quarters of all race horses suffer from this ailment. While EIPH is generally not fatal, it is known to reduce the performance of the horse. They are often slower to finish (not ideal for a race horse) and their racing lifespan can be affected. EIPH is usually measured using endoscopic methods and reported on a scale of 0 (no blood in airway) to 4 (airways full of blood), as an ordinal response.

In an effort to understand what some of the contributing factors could be for EIPH, Dr. Warwick Bayley and his colleagues at Washington State University's College of Veterinary Medicine took data on horses after a series of races. A part of this data was shared with us for the purpose of illustrations in the book. The data that we use is part of a much larger dataset and several manuscripts are in progress at the time of this book regarding various facets of the data. We use the data for illustration purposes only and are not answering the real scientific questions that arose from the data.

In the Horse Dataset we use in this book, we are given EIPH scores done by three veterinarians after a race as well as a consensus EIPH score. All horses in this study were 2-year-old thoroughbreds, who are eligible for some of the more famous races (Ascot, Derby, etc.). For this book, the predictors we consider are: distance of the race in meters (continuous), finish time taken in seconds (continuous) by the horse, the horse's starting and finishing position (discrete), type of surface for the race (categorical: turf, dirt, all weather), weather (categorical: clear, cloudy, rainy), and whether they were given a preventive medication Lasix (categorical: yes, no). Various other questions were asked for the research including the over-arching question, how can we model and potentially predict this type of data. We are using only a few explanatory variables for illustration out of a much bigger list.

1.2.2 Heart Data: Coronary Atherosclerosis

In the mid-1990s a very large-scale study was undertaken in Spokane, WA, USA to look at various aspects of health for people in high-risk professions. In this large multiple year study, coronary artery calcium, a measure of atherosclerosis, was measured in healthy adults along with a slew of predictors which included data related to life style, blood tests, food frequency, etc. One author of this book was involved in parts of the study. For this example, we are taking a very small part of the study for illustration. Here, the ordinal response is measured by a risk factor which was 0 (no risk), 1 (low risk), 2 (medium risk), and 3 (high risk) along with a few of the blood-related explanatory variables like glucose, gender, Body Mass Index (BMI), total cholesterol, cholesterol ratio (total/LDL), Red Blood Cells, White Blood Cells, potassium, diastolic blood pressure, systolic blood pressure, lactate dehydrogenase (LDH), and mean corpuscular hemoglobin (MCH). Like the Horse data, this dataset is used for illustration purposes only and is part of a much larger dataset. In this book we analyze this dataset of 691 participants and try to understand which of the twelve predictors contributed to a higher risk of coronary calcification.

1.3 Questions Arising in Ordinal Data

Ordinal data has often been thought of in the context of a tolerance model. The idea is that there is a latent variable in the background that is continuous in nature. The various ordered categories arise from some unobserved cut-points. In this book, we discuss the *commonly* used methods for ordinal data: Likert, Cumulative Distribution, Latent Variable and rank-based (nonparametric) methods. We discuss the intuition and mathematics behind these methods and use our datasets to illustrate the solutions.

The book is written as a data story. In the first section we provide more details of the datasets and the problems therein. In the second section, we take a closer look at the data in terms of summarizing the data, both graphically and numerically. The third section discusses the models that are used in the analysis of ordinal data and we will talk about the models in the context of the datasets and how they would be relevant for use (or not) depending upon the underlying questions using inference methods. This part of the book analyzes the data and looks for answers. We follow this up with diagnostics. Looking at what questions our analysis could answer and how appropriate are the results in the context of how closely were model assumptions satisfied. As most inferential research uses simulated data to ratify the techniques, we have included a chapter on simulating ordinal data. In the last section of the book we talk about the complete journey: from start to finish of the data story. To aid the thought progression this book is divided into five parts: Introduction, Exploratory Analysis, Methods of Analysis for ordinal data, Model Diagnostics, Simulation, Data Story and an Overall Summary. In the next sections we will talk briefly about R and SAS, the two software we used for analysis throughout the book.

1.4 Software Used: R and SAS

R and R Studio

To run the code in this book, you need R, R Studio, and some R packages (including ggplot2, MASS, and dplyr).

R is a free statistical computing language and environment that includes many modeling and graphical techniques. For this book, version 4.2.1 was used. However, there are several yearly updates for R and we recommend using the most updated version.

R Studio is a free, open source, integrated development environment (IDE) that uses R. R Studio's interface has 4 quadrants where you can easily write code and view graphs, output, datasets, files, code, objects, and more simultaneously. For this book, version 2023.9.0 was used. However, like with R, we recommend using the most updated version to ensure the functionality of the program.

To do the types of analyses in this book, additional packages beyond base R are needed. These packages can easily be installed using the "Packages" pane in R Studio or by running the line of code "install.packages ('name_of_package')" in the console. When a specific package is used, the code "library(name_of_package)" needs to be run in order to initialize the package. In this book, whenever a package is needed to run code, the library will be initialized at the start of the code chunk.

SAS and SAS Studio

SAS is one of the older statistical software. It was started in 1966 and then released to public in 1971. Unlike R it is not freely available for download. Because of that there is a fair amount of Research and Development and quality control in the processes run in SAS. From 2014 SAS has provided a free version for academics, which is used in this book. We have used SAS ON DEMAND FOR ACADEMICS to run all the code used in this book. SAS Studio 3.82 is compatible with SAS® 9.4M8. Some more information about the requirements for the environment is provided at: https://support.sas.com/en/documentation/systemrequirements.html. SAS has a variety of products each with a multitude of PROCEDURES (called PROC) in SAS. Most of the PROCs used in this book are from SAS/STAT.

Why R and SAS

We include code in both R and SAS since these are the most common programs used by the statistical community. We have tried to do the same analysis in both software for readers to get the logic and the syntax. We provide output as is given by the software (without editing) as this will allow readers to follow along with us seamlessly.

Part II

Exploratory Analysis

2

Summarizing and Visualizing Ordinal Data

2.1 Introduction

In this book, we focus on telling stories with data. By telling the story behind the data, we can contextualize the problem and come up with appropriate solutions. This process of storytelling begins in the Exploratory Data Analysis (EDA) stage where patterns, trends, and relationships (or lack thereof) are searched for prior to modelling the data. Then, summaries and visualizations again become advantageous when interpreting some model results and presenting final results to different audiences. So, this is a very important piece in how we analyze the data. In this chapter, we use the Horse dataset to demonstrate ways of analyzing ordinal data. We discuss summarizing ordinal data numerically and graphically. While we attempt as much as possible to use questions that are related to the dataset, we sometimes discuss generic cases that are appropriate for ordinal data even though they are not appropriate for the example dataset.

Re-visiting the Horse Dataset

In the last chapter, we described the Horse Dataset which consists of data related to hemorrhaging after strenuous exercise. For each horse, we have the EIPH score which is our response variable, and seven explanatory variables of interest. These are (1) distance run in the race, (2) time of completion, (3) starting position, (4) finishing position, (5) type of surface, (6) weather condition and (7) treatment type for EIPH before the race, and (8) speed index.

In this chapter, for each graphical or numerical summary, we will provide the code and output from R. We will also provide the relevant code for SAS.

2.2 Graphical Summary

The tricky part of summarizing ordinal data is the ambiguity of whether the ordinal response is numerical or categorical. For categorical data, pie charts

DOI: 10.1201/9781003020615-2

and barplots are common methods of visualization because these plots are great at visualizing frequency counts of groups. These plots also work well for visualizing ordinal data. However, due to the fact that the categories are ordered for ordinal data, other types of visualizations including histograms, step diagrams, boxplots and scatter plots are also possible. We will pursue both types of graphical summaries.

2.2.1 Univariate Ordinal Data

A crucial first step in Exploratory Data Analysis (EDA) is to visualize the response variable. This allows us to get an idea of the general shape of the data. Any method for summarizing ranks is an option for univariate ordinal data. In Figure 2.1(a-d), we provide various summaries for the response variable, Consensus EIPH. Consensus EIPH is the score that was agreed upon by three veterinarians after they had individually scored the horses. The possible values of Consensus EIPH are 0 (absence of any blood in the lungs), 1, 2, 3, and 4 (lungs filled with blood). In each case, we see that for each rank we have a proportion of horses. Hence, this variable structure makes it suitable for a barplot like what is used for categorical data. In the case of an ordinal variable, a barplot is identical to a histogram which allows us to talk about the shape of the distribution in terms of symmetry and skewness. Other ways to visualize and summarize the ordinal response includes pie charts and frequency tables.

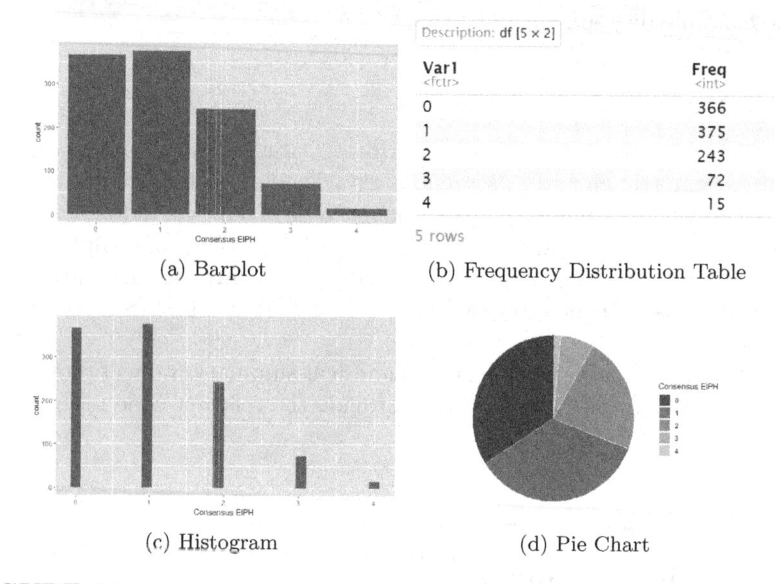

Var1 <fctr>	Freq <int>
0	366
1	375
2	243
3	72
4	15

(a) Barplot

(b) Frequency Distribution Table

(c) Histogram

(d) Pie Chart

FIGURE 2.1
Visualizations and summaries for univariate ordinal data

Plot (a) of Figure 2.1 presents a barplot that displays the frequencies of the Consensus EIPH levels. Consensus EIPH 4 has the lowest frequency (15 horses) while Consensus EIPH 1 has the highest (375 horses). The variable is right skewed, which means that most of the horses were on the lower end of the Consensus EIPH scale. Plot (d) represents a pie chart that tells a similar story like the barplot in Plot (a), but with the ability to compare the frequencies proportionally. Since the ordinal variable is correctly ordered in the plot, we can read the pattern similarly to reading the barplot. Plot (b) is a table of frequencies that reports the number of horses for each Consensus EIPH level. We observe that 366 horses (34.17% of the total) have an Consensus EIPH of level 0, and 375 horses (35.01% of the total) have a Consensus EIPH of level 1. In the middle, 243 horses (22.69% of the total) have a Consensus EIPH level of 2. For the highest levels of Consensus EIPH (3 and 4), we count respectively 72 horses (6.72% of the total) and 15 horses (1.4% of the total).

Overall, the graphical summaries indicate that Consensus EIPH is generally skewed in shape, with larger percentages of horses having no bleeding or little bleeding in the lungs. This is potentially good news for the racing community as higher Consensus EIPH is a detriment to racing trophies. The R Code for all four graphs are provided. The code to import a dataset is in the section at the end of this chapter.

Below is the code to create the plots and table. To remove the legend in the visualizations, the following layer is used: '+ theme(legend.position = "none")'.

```
1  library(ggplot2) ## loading the library ggplot2
2  library(dplyr) ## loading the library dplyr
3
4  # Barplot
5  DATA %>%
6    ggplot(aes(x = ConsensusEIPH, fill= ConsensusEIPH))
     +
7    geom_bar(stat = "count")+
8    xlab("Consensus EIPH")+
9    theme(legend.position = "none")
10
11 # Frequency distribution tables
12 as.data.frame(table(DATA$ConsensusEIPH))
13
14 # Histogram
15 DATA %>%
16   ggplot(aes(x = ConsensusEIPH, fill = ConsensusEIPH)
     )+
17   geom_bar(stat = "count")+
18   xlab("Consensus EIPH")+
```

```
  theme(legend.position = "none")

# Pie Chart
DATA_table=as.data.frame(table(DATA$ConsensusEIPH))

DATA_table %>%
  ggplot(aes(x = "", y = Freq, fill = Var1)) +
  geom_bar(stat = "identity", width = 1, color = "
    white") +
  coord_polar("y", start = 0)+
  guides(fill = guide_legend(title = "Consensus EIPH"
    ))+
  theme_void()
```

The relevant but generic SAS code for frequency tables, pie chart, and barplot are given below. Here our dataset is called "dataset" containing the ordinal response Y.

```
/* barplot and pie chart */
Y= ordinal response

proc gchart data=dataset ;
pie Y;
hbar Y;
run;

/* Frequency Table */
proc freq data=dataset;
table Y ;
run;
```

After looking at the response variable in terms of its shape and structure, the next step is to look at how it relates to the different explanatory variables.

2.2.2 Two Variables: Ordinal Response and Categorical Explanatory Variable

We investigate the two-way patterns between the ordinal variable and a categorical variable. To do this, the ordinal response will be treated as ordered categorical and the categorical explanatory variable will be treated as categorical.

In the table and plots, we explore the response variable, Consensus EIPH, across the weather, surface and Lasix variables.

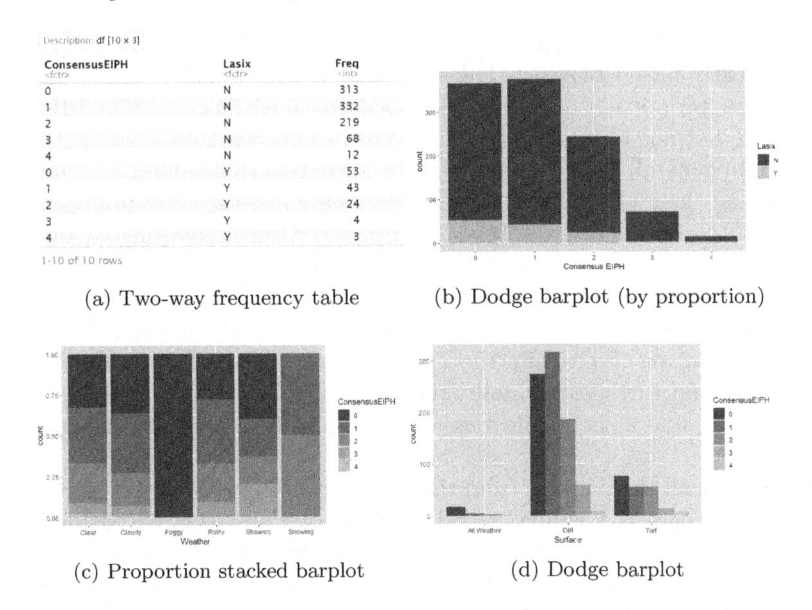

Description: df [10 × 3]

ConsensusEIPH <fctr>	Lasix <fctr>	Freq <int>
0	N	313
1	N	332
2	N	219
3	N	68
4	N	12
0	Y	53
1	Y	43
2	Y	24
3	Y	4
4	Y	3

1-10 of 10 rows

(a) Two-way frequency table

(b) Dodge barplot (by proportion)

(c) Proportion stacked barplot

(d) Dodge barplot

FIGURE 2.2
Two variables: Ordinal response and categorical explanatory variable

Plot (a) of Figure 2.2 is a cross-table (or contingency table) that displays the frequencies of Consensus EIPH Level by Lasix treatment. Lasix is a dichotomous indicator taking two possible responses, yes or no, where yes means the horse was under this type of treatment and no refers to no treatment. For example, among the total of 366 horses of 0 Consensus EIPH level, 313 did not receive the Lasix while 53 received this treatment. For the total of 15 horses exposed to the most severe category of Consensus EIPH (level 4), 12 did not receive Lasix while 3 received the Lasix treatment. Plot (b)) explains the same story, but with a dodge barplot (by proportion). For instance, the bar of Consensus EIPH level 2 indicates that 219 horses received no Lasix while only 24 horses received this treatment. Overall, no Lasix treatment was prevalent for each Consensus EIPH category. However, the results appear this way because there are overall more horses in the dataset that have had the Lasix procedure. A better way to explore these variables would be to use a proportion stacked barplot where the relationships are totally assessed using proportions (for example in Plot (c)). When a proportion stacked barplot is used for these variables (not pictured), it is easier to see that the proportions of horses that have had the Lasix procedure are similar across the different Consensus EIPH levels.

Plot (c) of Figure 2.2 is a stacked proportion barplot that shows the Consensus EIPH level by different weather conditions (clear, cloudy, foggy, rainy, showery, and snowing). For snowy weather, we see that the horses competing all had a Consensus EIPH of 1 or 2, while for foggy weather, the horses that competed had Consensus EIPH of 0. In the other weather types, the horses mostly have Consensus EIPH of 0, 1 and 2, but snowing weather had the highest proportion of horses with Consensus EIPH of 3. Plot (d) shows the dodge barplot to explore the relationship between Consensus EIPH and Surface. In general, there were more data for horses that ran on dirt, followed by turf, and then followed by All weather. This can be seen by comparing the heights of the bars across the types of surfaces. After accounting for numbers of horses that ran on each surface, the patterns of Consensus EIPH for the different surfaces can be discussed. For example, we see that more horses had Consensus EIPH of 1 as compared to horses that had Consensus EIPH of 0 when they ran on dirt, while the opposite was true for horses that ran on all weather and turf. Additionally, the number of horses that had Consensus EIPH of 2, 3, and 4 show a decreasing trend for all three types of surfaces.

Summarizing all the findings from the different visualizations and summaries, we see that there is a similar trend in levels of Consensus EIPH for horses who had the Lasix treatment versus those that did not have the treatment. In terms of the weather, Consensus EIPH had a similar trend in clear and cloudy weather and Consensus EIPH was the highest in rainy and snowing weather. For surface, the only difference in trend was that more horses had Consensus EIPH of 1 compared to 0 when they ran on dirt and the opposite was true when the horses ran on all weather surface and turf.

The R Code is provided below.

```
library(ggplot2) ## loading the library ggplot2
library(dplyr) ## loading the library dplyr

# Two-way frequency table
table = as.data.frame(table(DATA$ConsensusEIPH, DATA$
   Lasix))
colnames(table) = c("ConsensusEIPH","Lasix", "Freq")

# Dodge barplot (by proportion)
 DATA %>%
  ggplot(aes(x = ConsensusEIPH, fill = Lasix))+
  geom_bar()+
  xlab("Lasix")


# Proportion stacked barplot
 DATA %>%
  ggplot(aes(fill = ConsensusEIPH, x=Weather))+
  geom_bar(position = "fill")+
  xlab("Weather")

```

```
# Dodge barplot
DATA %>%
  ggplot(aes(fill = ConsensusEIPH, x=Surface))+
  geom_bar(position = "dodge")+
  xlab("Surface")
```

The SAS code for two-way frequency tables and barplots charts is provided below. Here our dataset is called "dataset" containing the ordinal response Y and the explanatory variable X1.

```
/*Two-way frequency table*/
PROC FREQ data=dataset;
TABLE Y*X1;
run
/*  barplots */
PROC GCHART data=dataset;
HBAR Y/subgroup=X1;
run;
```

2.2.3 Two Variables: Ordinal Response and Ordinal Explanatory Variable

As both the response and explanatory variables are ordered, the types of plots and summaries from Figure 2.2 can be used if the ordinal response is treated as ordered categorical and the ordinal explanatory variable is treated as ordered categorical. In addition, boxplots and violin plots can be used if the ordinal response is treated as ordered categorical and the ordinal explanatory variable is treated as numeric.

In the plots below, we demonstrate the visualizations not used previously. We explore the response variable, Consensus EIPH, across the finish position variable (Finish).

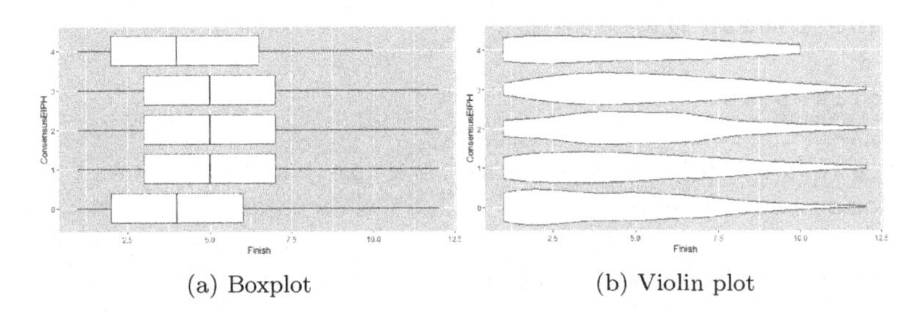

(a) Boxplot (b) Violin plot

FIGURE 2.3
Two variables: Ordinal response and ordinal explanatory variable

Plot (a) of Figure 2.3 is a boxplot that displays side-by-side box-and-whiskers of Consensus EIPH grouped by finish positions. The ends of the boxes—showing group 25th and the 75th percentiles are near the lower end of the distribution for the finish positions. These suggest a right-skewed distribution for each finish position. The lines in the middle of the boxes—showing the medians, suggest that the medians for Finish positions for the horses that had Consensus EIPH of 0 and 4 were similar while the horses that had consensus EIPH of 1, 2, and 3 were similar. The medians for the horses with Consensus EIPH of 0 and 4 were larger than the medians ot the horses that had Consensus EIPH of 1, 2, and 3. We observe no horses that were considered as outliers.

Plot (b) of Figure 2.3 presents the same data but with a violin plot of the Consensus EIPH. A violin plot is a hybrid of the traditional boxplot and kernel density plots. So, we get the additional insight of the density of horses that had a specific finish position within a certain Consensus EIPH. For example, horses that had a Consensus EIPH of 0 had the most right skewed distribution of finish positions which is evidenced by the thickest violin near the right end of the distribution of finish place. Practically, this means that majority of the horses that had a Consensus EIPH of 0 had a lower finish position as compared to the other horses that had other Consensus EIPH's.

The R Code is provided below. To make the plots, we have to treat the ordinal explanatory variable as numerical. This allows us to see the distribution of that variable.

```
1  library(ggplot2) ## loading the library ggplot2
2  library(dplyr) ## loading the library dplyr
3
4  # Boxplot
5  DATA%>%
6    mutate(ConsensusEIPH = as.factor(as.character(
       ConsensusEIPH)),
7                  Finish = as.numeric(Finish))%>%
8    ggplot(aes(x = Finish, y = ConsensusEIPH))+
9    geom_boxplot()+
10   xlab("Finish")
11
12 # Violin plot
13 DATA %>%
14   mutate(ConsensusEIPH = as.factor(as.character(
       ConsensusEIPH)),
15                 Finish = as.numeric(Finish))%>%
16   ggplot(aes(x = Finish, y = ConsensusEIPH))+
17   geom_violin()+
18   xlab("Finish")
```

The generic SAS code for boxplots is provided below. Here our dataset is called "dataset" containing the ordinal response Y and the explanatory variable X1

```
/* Boxplot */
proc boxplot ;
 plot Y*X1;
 run ;
```

2.2.4 Two Variables: Ordinal Response and Discrete Numerical Explanatory Variable

To explore the relationships between an ordinal response and a discrete numerical explanatory variable, the types of boxplots and violin plots illustrated in Figure 2.3 can be used. In this case, the ordinal response is treated as ordered categorical and the discrete numerical explanatory variable is treated as ordered numeric. In addition, we may use scatter plots but with the caveat that the ordinal response is treated as a numeric and the discrete numerical explanatory variable is treated as numeric.

Consider the plot below, showing how the response variable and Consensus EIPH, are related to the speed index variable. Since this is a scatterplot, we include a LOESS line to help determine if there is a linear relationship between the two variables (or lack thereof). A LOESS line is generated using a nonparametric method that uses local regression to fit a smooth curve through a scatterplot of data. Plots with a LOESS line allows us to see what scatter plots often cannot show us for ordinal responses—the type of relationship patterns.

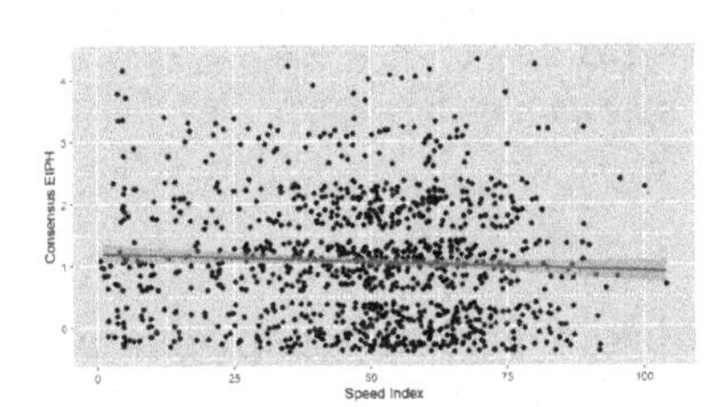

FIGURE 2.4
Two variables: Ordinal response and Discrete Numerical explanatory variable

Plot (a) of Figure 2.4 presents a scatterplot of Consensus EIPH and speed index. The distribution of the points suggest that most of the horses of Consensus EIPH levels 0, 1, and 2 show that speed index concentrated between 30 and 80 seconds. There are a lot less horses with Consensus EIPH of 3 and 4, but horses with Consensus EIPH 3 had speed index on a much wider range than horses with Consensus EIPH 4. Otherwise, there does not seem to be a trend when looking at the data points even though the LOESS line appears to show a slightly negative or decreasing trend.

The R Code is provided below. To make the plot, we have to treat both the ordinal explanatory variable and the ordinal response variable as numerical. This plot has the points jittered, so that the plot is easier to read. Without jittering, all the points with the same discrete value would lie on top of each other and that makes it difficult to see the number of points around the same region geom_jitter() is used to jitter the data points and geom_smooth() is used to add the LOESS line.

```
library(ggplot2) ## loading the library ggplot2
library(dplyr) ## loading the library dplyr

# Scatterplot with LOESS line
DATA %>%
  mutate(ConsensusEIPH = as.numeric(as.character(
    ConsensusEIPH)),
                  'Speed Index' = as.numeric('Speed
    Index'))%>%
  ggplot(aes(x= 'Speed Index', y = ConsensusEIPH ))+
  geom_jitter()
  + geom_smooth()
  ylab("Consensus EIPH")
```

The SAS code for scatter plot with loess line is provided below. Here our dataset is called "dataset" containing the ordinal response Y and the explanatory variable X1. The LOESS smoothing parameter is 0.5.

```
/*scatter plot with loess line*/
proc sgplot data=dataset ;
loess x=X1 y=Y/ smooth=.5;
run;
```

2.2.5 Two Variables: Ordinal Response and Continuous Explanatory Variable

To see the relationships between an ordinal response and a continuous explanatory variable, we can again use both boxplots, violin plots, and scatterplots. For a boxplot or violin plot, the ordinal response will be treated as a ordered categorical, and the continuous explanatory variable will be treated as numeric. For a scatterplot, the ordinal response will be treated as a numeric, and the continuous explanatory variable will be treated as numeric.

Consider the following plots, showing how the response variable and Consensus EIPH, are related to finish time.

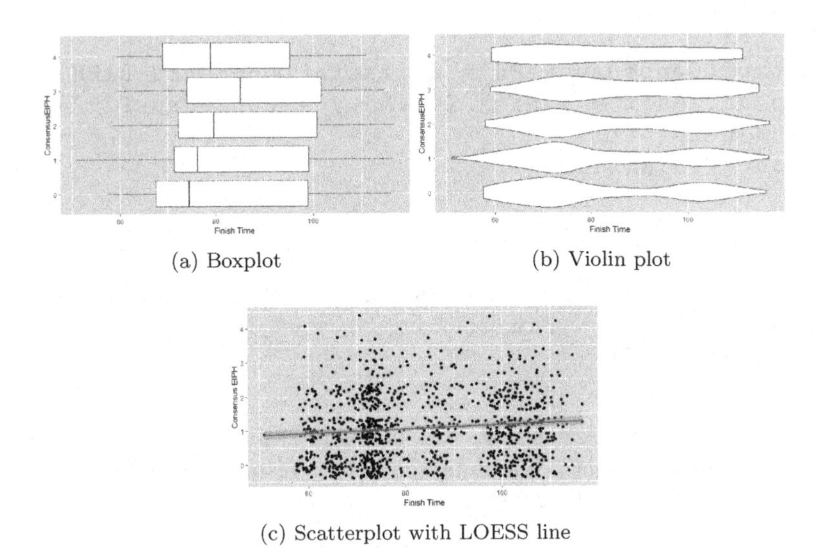

(a) Boxplot

(b) Violin plot

(c) Scatterplot with LOESS line

FIGURE 2.5
Two variables: Ordinal response and continuous explanatory variable

The plots in Figure 2.5 do not show a trend. When looking at the scatterplot in plot (c), the points are spread out. There is not a clear linear or non-linear trend of Consensus EIPH based on Finish time, even though the LOESS line suggests that there is a slightly increasing or positive trend. Likewise, the boxplot and violin plot (Plots (a) and (b)) show that the distributions of Finish Time is spread out over the entire range of Finish times regardless of the Consensus EIPH group. From the violin plot, the Finish times seem to be concentrated in the lower and upper ends of the distribution (based on the width of the violins) for Consensus EIPH of 0, 1, 2, and 3. There were very little data for horses with Consensus EIPH 4, so that trend is not being interpreted here.

The R Code is provided below.

```
1 library(ggplot2) ## loading the library ggplot2
2 library(dplyr) ## loading the library dplyr
3
4 # Scatterplot
5 DATA %>%
6   mutate(ConsensusEIPH = as.numeric(as.character(
      ConsensusEIPH)))%>%
7   ggplot(aes(x = ‘Finish Time‘, y = ConsensusEIPH))+
```

```
  geom_jitter()+
  xlab("Finish Time")

# Boxplot
DATA %>%
  mutate(ConsensusEIPH = as.factor(ConsensusEIPH))%>%
  ggplot(aes(x =‘Finish Time‘, y = ConsensusEIPH))+
  geom_boxplot()+
  xlab("Finish Time")

# Violin plot
DATA %>%
  mutate(ConsensusEIPH = as.factor(ConsensusEIPH))%>%
  ggplot(aes(x = ‘Finish Time‘, y = ConsensusEIPH))+
  geom_violin()+
  xlab("Finish Time")
```

The SAS code for scatterplots and boxplots is provided below. Here our dataset is called "dataset" containing the ordinal response Y and the explanatory variable X1. The LOESS smoothing parameter is 0.5.

```
/*Scatterplot*/
PROC SGPLOT data=dataset ;
LOESS x=X1 y=Y/ smooth=.5;
run;

/*Boxplot*/
PROC BOXPLOT data=dataset;
PLOT Y*X1;
run;
```

2.3 Numerical Summary

With numerical summaries, we look at measures of central tendency or spread, rather than graphical summary. Hence, even if one is on the fence about the "numeracy" of ordinal data, a measurement like *Mode* indicating the most common category would be an appropriate *numerical* summary of the data. Frequency measures, which is often used for categorical data, is also an appropriate numerical summary.

However, whatever ones' opinions about ordinal data are, it is clear that the response variable can be treated as a *rank* and we can use summarizing methods for ranks. As a result, we can use measures like mean ranks and standard deviation of ranks in that context. Median ranks as well as Median Absolute Deviation (MAD) measures can be used as well.

2.3.1 Univariate Ordinal Data

Any method for summarizing ranks is an option for univariate ordinal data. We provide the frequency counts for Consensus EIPH. We also provide the mean, median, mode, MAD, and standard deviation of the response variable.

Consider the tables below showing summaries of the response variable.

mean <dbl>	median <dbl>	mode <dbl>	sd <dbl>	MAD <dbl>
1.061625	1	1	0.9801453	1.4826

1 row

(a) Numerical summary of the ordinal response (Consensus EIPH)

0	1	2	3	4
366	375	243	72	15

(b) Frequency counts for the ordinal response (Consensus EIPH)

FIGURE 2.6
Univariate ordinal data

The R Code is provided below.

```
library(dplyr) ## loading the library dplyr

# Function to calculate the mode since this does not
   yet exist in R
mode <- function(m) {
   uniquem <- unique(m)
   uniquem[which.max(tabulate(match(m, uniquem)))]
}

# Numerical summary of the ordinal response
DATA %>%
    mutate(ConsensusEIPH = as.numeric(Consensus EIPH)
   )%>%
    summarize(mean = mean(ConsensusEIPH, na.rm= TRUE)
   ,
            median = median (ConsensusEIPH, na.rm=
   TRUE),
            mode = mode(ConsensusEIPH),
            sd = sd(ConsensusEIPH, na.rm=TRUE),
```

```
                  MAD = mad(ConsensusEIPH, na.rm=TRUE))

# Frequency counts for the ordinal response
ftable(DATA$ConsensusEIPH)
```

The SAS code for frequency tables and numerical summaries is provided below. Here our dataset is called "dataset" containing the ordinal response Y and the explanatory variable X1.

```
/*Frequency counts for the ordinal response*/

PROC FREQ;
TABLE Y*X1;
run;

/*Numerical summary of the ordinal response*/

PROC MEANS;
VAR Y;
run;
```

2.3.2 Two Variables: Ordinal Response and Categorical Explanatory Variable

We tabulate the frequency of the responses by the categorical explanatory variables. This is the same as constructing two-way contingency tables to see how the variables are related. Similarly we estimate the mean, median, and mode by the categorical variable. This gives us an understanding of how the response variable varies across the categorical variables.

Consider the tables below showing how the response variable is related to weather.

Table (a) of Figure 2.7 presents the numerical summary of Consensus EIPH based on different weather conditions. We observe that for all weather types, the mean Consensus EIPH level was about the same as the median and the mode. Overall, this table shows that the Consensus EIPH level can vary depending on the weather condition, with most conditions (clear, cloudy, rainy, and showery) associated with lower Consensus EIPH levels than snowing. However, it is important to note that this is a descriptive analysis; thus it does not establish any causal relationship between the weather and the Consensus EIPH level.

Table (b) of Figure 2.7 is a contingency table which shows the number of horses for each combination of the Consensus EIPH level (0 to 4) and weather conditions (clear, cloudy, foggy, rainy, showery, and snowing). For instance, there are 180 horses in the group of horses with Consensus EIPH of 0 that

A tibble: 6 x 4

Weather <chr>	mean <dbl>	median <dbl>	mode <dbl>
Clear	1.109091	1.0	1
Cloudy	0.988764	1.0	1
Foggy	0.000000	0.0	0
Rainy	1.139535	1.0	1
Showery	1.166667	1.0	0
Snowing	1.500000	1.5	2

6 rows

(a) Numerical summary for ordinal response (Consensus EIPH) by the categorical variable (weather)

	Clear	Cloudy	Foggy	Rainy	Showery	Snowing
0	180	161	1	12	12	0
1	187	163	0	17	7	1
2	136	91	0	10	5	1
3	37	25	0	4	6	0
4	10	5	0	0	0	0

(b) Two way contingency table of the ordinal response (Consensus EIPH) and a categorical factor (weather)

FIGURE 2.7
Two variables: Ordinal response and categorical factor

performed under clear weather conditions. Similarly, there are 187 horses with Consensus EIPH level 1 that performed under clear weather condition. Two-thirds (10) of the 15 horses with highest level of Consensus EIPH performed under clear weather conditions, while the other one-third performed under cloudy weather conditions. These are a few examples of how this table can be used to examine the distribution of Consensus EIPH levels across different weather conditions. Overall, we can see that Consensus EIPH has a similar trend when horses performed under most weather conditions.

The R Code is provided below.

```
library(dplyr) ## loading the library dplyr

# Numerical Summary for ordinal response by the
  categorical variable
DATA  %>%
    mutate(ConsensusEIPH = as.numeric(ConsensusEIPH))
  %>%
    group_by(Weather)%>%
    summarize(mean = mean(ConsensusEIPH, na.rm= TRUE)
  ,
            median = median (ConsensusEIPH, na.rm=
  TRUE),
            mode = mode(ConsensusEIPH))

```

```
# Two way contingency table of the ordinal response
    by the categorical variable
ftable(DATA$ConsensusEIPH, DATA$Weather)
```

The generic SAS code for summary is provided below. Here our dataset is called “dataset” containing the ordinal response Y and the explanatory variables X1, X2 and X3.

```
/* Numerical Summary for ordinal response by the
    categorical variable */
PROC MEANS data=dataset;
VAR Y X1 X2 X3;
run;
```

2.3.3 Two Variables: Ordinal Response and Ordinal Explanatory Variable

Similar to using a categorical explanatory variable, we tabulate the frequency of the responses by the ordinal explanatory variable to see how the variables are related. Likewise, we estimate the mean, median, and mode of the ordinal response by the ordinal explanatory variable. This gives us an understanding of how the response varies across the ordinal variables. Here, if both variables are ranks, we can calculate a correlation coefficient between the ranks.

Consider the tables below showing how the response variable is related to Finish and Starting Place.

Table (a) of Figure 2.8 is a table of frequency which presents the frequency distribution of Consensus EIPH based on the finish position of the horses in the race. The finish position ranges from 1 to 12, with each row representing a different level of Consensus EIPH, from 0 to 4. For example, the first row shows that 54 horses of Consensus EIPH level 0 finished in first position, while 42 horses with an Consensus EIPH of level 1 finished in the first position. The frequency of horses with Consensus EIPH level 0 decreased gradually as the finish position increased, with only one horse finishing in position 12. Horses with an Consensus EIPH level 1 were most frequent at finish positions 2 to 5. Similar to Consensus EIPH of 0, the frequency of horses with Consensus EIPH level 2 decreased as the finishing position increased.

Table (b) of Figure 2.8 presents the summary statistics of the Consensus EIPH level by the starting place of the horses. The mean indicates the average Consensus EIPH level of horses starting at each position; the median shows the middle Consensus EIPH level among horses starting at each position; and the mode shows the most frequent Consensus EIPH level among horses starting at each position. We observe that as starting place increases, the mean Consensus EIPH also increases. The median Consensus EIPH was the same for

	1	2	3	4	5	6	7	8	9	10	11	12
0	54	59	49	37	49	34	37	19	14	8	5	1
1	42	47	49	49	44	37	33	30	15	17	7	5
2	24	16	29	39	29	35	29	11	15	7	7	2
3	2	9	8	12	8	9	7	7	4	4	1	1
4	3	2	2	1	1	2	2	0	1	1	0	0

(a) Frequency of responses by ordinal variables (Consensus EIPH by Finish Place)

A tibble: 13 x 4

Starting Place <dbl>	mean <dbl>	median <dbl>	mode <dbl>
1	0.8521739	1	0
2	0.9292035	1	0
3	0.9779412	1	0
4	1.0135135	1	0
5	0.9926471	1	1
6	1.1475410	1	1
7	1.2735849	1	1
8	1.1600000	1	1
9	1.2400000	1	1
10	1.1315789	1	1

1-10 of 13 rows Previous 1 2 Next

(b) Numerical summary of Consensus EIPH by Ordinal explanatory variable (Starting Place) - first 10 rows presented

FIGURE 2.8
Two variables: Ordinal response and ordinal explanatory variable

all starting positions. However, the mode Consensus EIPH was lower for the lower starting positions than the higher starting positions.

Overall, the results suggest that there is a relationship between the start and finish place of horses and the likelihood of experiencing different levels of Consensus EIPH. Horses starting from lower positions are less likely to experience higher Consensus EIPH than horses starting from higher positions. The opposite trend seems to be true for finish position.

The R Code is given below.

```
1  #Frequency of responses by ordinal variables
2  table(DATA$ConsensusEIPH, DATA$starting)
3
4  #Numerical Summary by Ordinal explanatory variable
5  DATA %>%
6    mutate(ConsensusEIPH = as.numeric(ConsensusEIPH))
     %>%
7    group_by(`Starting Place`)%>%
8    summarize(mean = mean(ConsensusEIPH, na.rm= TRUE),
9              median = median(ConsensusEIPH, na.rm=
     TRUE),
10             mode = mode(ConsensusEIPH))
```

2.3.4 Two Variables: Ordinal Response and Discrete Numerical Explanatory Variable

All the techniques used in the last subsection is relevant for this section. If the number of discrete outcomes are not too large, we can construct two way contingency tables. However, we can construct numerical summary tables regardless of the number of discrete outcomes. For this example, we demonstrate the correlation coefficient between starting place and Consensus EIPH.

```
[1] 0.1382508
```

Correlation coefficient

FIGURE 2.9
Two variables: Ordinal response and discrete numerical explanatory variable

The R Code is provided below.

```
1 library(dplyr) ## loading the library dplyr
2
3 # The response needs to be numeric for this analysis
4 # This updates the response variable to be numeric
5 DATA_response_numeric = DATA %>%
6                           mutate(‘ConsensusEIPH‘ = as
    .numeric(ConsensusEIPH))
7
8 #Correlation coefficient
9 cor(DATA_response_numeric$ConsensusEIPH, DATA_
    response_numeric$‘Starting Place‘, method = "
    pearson", use = "complete.obs")
```

The generic SAS code for correlations is provided below. Here our dataset is called “dataset” containing the ordinal response Y and the explanatory variables X1, X2, and X3.

```
1  /* Correlation Coefficient */
2 PROC CORR data = dataset ;
3 VAR Y X1 X2 X3;
4 run ;
```

This result is interpreted as there being a weak linear correlation between Consensus EIPH and starting place. This does not imply that there is not a non-linear relationship between the variables.

2.3.5 Two Variables: Ordinal Response and Continuous Explanatory Variable

Again, we can calculate a correlation coefficient to see if there is a linear relationship. Ordinary Least Square (OLS) analysis could be done as an exploratory method. However, OLS has to be interpreted with much care.

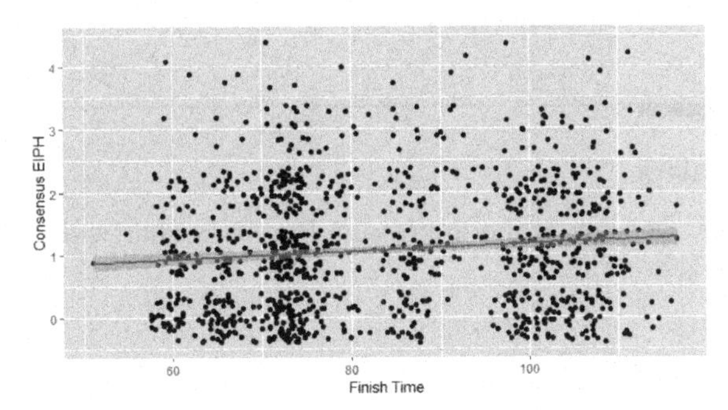

(a) Scatterplot with estimated linear regression line

```
[1] 0.1017138
```

(b) Correlation coefficient

FIGURE 2.10
Two variables: Ordinal response and discrete Numerical explanatory variable

The R Code is provided below.

```
# Scatterplot with estimated linear regression line
DATA %>%
  ggplot(aes(x=ConsensusEIPH, y= 'Finish Time'))+
  geom_jitter()+
  geom_smooth(method=lm)+
  xlab(" Consensus EIPH")


# Correlation coefficient

# The response needs to be numeric for this analysis
# This updates the response variable to be numeric
DATA_response_numeric = DATA %>%mutate('ConsensusEIPH
    ' = as.numeric(ConsensusEIPH))

#Correlation coefficient calculation
cor(DATA_response_numeric$ConsensusEIPH, DATA_
```

```
response_numeric$'Finish Time', method = "pearson"
, use = "complete.obs")
```

Both the scatterplot with estimated linear regression line in Figure 2.10 (a) and the correlation coefficient in Figure 2.10 (b) show a weak linear correlation between Consensus EIPH and Finish time.

2.4 R Syntax

In this chapter, we show some important R syntax that will be used for the rest of the book.

Importing a dataset

To be able to use the dataset in R, we have to first import the dataset. Here are a few options:

```
# OPTION 1 for csv or tsv files

#The readr package is used to import csv and tsv
    files into R
library(readr)     # loading the readr package

# The general format of the code is:
# dataset_name_in_r = read_csv("name_of_file.csv")

# For the Horse race example used in the book, the
    code is:
DATA = read_csv("EIPHdata.csv")

# OPTION 2 for xlsx or xls files
library(readxl)  # loading the readxl package
dataset_name_in_r <- read_excel("name_of_file.xlsx")
dataset_name_in_r <- read_excel("name_of_file.xls")
```

Using pipes

Pipes (%>%)—a pipe is used to linearly apply a function to a dataset. For example, if you want to filter a dataset using a criteria, you can apply the filter function to the dataset using a pipe. The basic structure for using a pipe is:

```
library(dplyr) ## loading the library dplyr
dataset_name_in_r %>% function_to_apply_for_example_
    filter()
```

Creating graphs Using ggplot2 package

ggplot is a function that is used to create graphs. To create a graph, you define the dataset, the aesthetics (i.e., mapping of the variables to the different parts of the graphs), and add layers (using +) to add more features to the graph (for example, the type of graph desired, titles, change the color, legend, title, etc.). Below is the basic structure of a ggplot. In this example, we create a scatterplot with a linear regression trend line.

```
library(dplyr) ## loading the library dplyr

 dataset_name_in_r %>%
  ggplot(aes(x=variable_1 y= variable_2))+
  geom_jitter()+
  geom_smooth(method=lm)+
  xlab("Title for the x axis")+
  ylab("Title for the y axis")
```

In this case, there is a pipe that is used to apply the ggplot function to the dataset. Then, inside the ggplot function, we define the aesthetics (the x axis variable, y axis variable, etc.). To add additional features to the plot, there are multiple layers. The first layer requests a scatterplot (i.e., + geom_jitter). The next layer requests a trend line using a linear model (+ geom_smooth(method=lm)) and the last layer adds an x axis label + xlab ("Title for the x axis")). There are many other types of layers that can be added, but will not be discussed in this book.

Changing variable types

Variable types should always be coded correctly. Categorical and Ordinal variables should be coded as a FACTOR and numerical variables should be coded as NUMERIC. It is very easy to do this when the dataset is first imported into R. However, we treat the variables differently depending on the type of plot or table we create, and so this can be done before creating each plot or table.

To code a variable as factor:

```
# The dplyr package is used for many data
    manipulation tools
library(dplyr) # loading the dplyr package

 dataset_name_in_r = dataset_name_in_r %>% mutate(New
```

```
    Variable_name=as.factor(Variable_that_needs_to_be
    _changed))
```

To code a variable as numeric:

```

library(dplyr) # loading the dplyr package

dataset_name_in_r = dataset_name_in_r %>% mutate(New_
    Variable_name=as.numeric(Variable_that_needs_to_be
    _changed))
```

In the code above, we overwrite the original dataset with the new dataset that has the changed variables that are of appropriate variable type.

2.5 SAS Syntax

SAS syntax uses Procedures called "PROC" statements. The logic is, we invoke the Procedure and assign a dataset to it along with the options available. As we not not using color for the code, we are using CAPITAL letters for the inbuilt SAS words:

PROC "generic procedure" DATA= dataset;

Each SAS line ends with a semi-colon ";".
Once a procedure is invoked and the dataset is read in for the procedure, we will need to follow syntax for the procedure.

For example,

```
PROC MEANS DATA=dataset;
var Y X1;
run;
```

Here the term "VAR" refers to the variables that SAS will summarize. Each SAS snippet ends with a run statement.

Importing a dataset

Just as in other processes, SAS uses the Procedure IMPORT to import data. SAS can import data from Excel, csv, and a multitude of files, using PROC IMPORT.

```
FILENAME REFILE '/home/path/dataset.xlsx';

PROC IMPORT DATAFILE=REFFILE
  DBMS=XLSX
  OUT=dataname;
  GETNAMES=YES;
RUN;

PROC CONTENTS DATA=dataname; RUN;
```

This program will find the dataset given the path and import it as a SAS dataset. If we use a .csv file, then the line DBMS=XLSX will change to DBMS=CSV.

Some facts about SAS

- SAS is not case sensitive, so data is the same as DATA unlike R. Unless it is within inverted commas as then it is read as a name. So, "A" will be a different entity than "a".
- As R uses libraries, SAS has a multitude of PROC statements with its own syntax.
- SAS does not admit spaces between the name of a variable. It is best to say "cholratio" as opposed to "chol ratio".
- Preferably variable name should be kept to less than 8 characters.

2.6 Summary

In this chapter we went through the process of exploratory data analysis (EDA) where we summarize and visualize our data before analysis. This stage is crucial, as it allows us to see discrepancies in the data type, missing observations, and wrong syntax of data. The graphs also allow us to see patterns and outliers. With the advent of computing and software, EDA is a required first step as it helps us catch issues before we use computationally intensive methods to analyze data.

Part III

Methods for the Analysis of Ordinal Data

3

Historical Perspective

After summarizing data using graphs and numerical summaries, we have a general idea about the shape of the response variable, the types of relationships that exist (or not) within the dataset, and the issues with the data. We do want to emphasize that the questions asked about the data are done prehoc. So, while the EDA done in chapter 2 gives us an idea of patterns and shapes, the plan of analysis is based on the questions asked before the data was collected.

However, before we delve into analyzing ordinal data, we want to pause and look at the various ways one *could* analyze ordinal data and discuss their background and history. First, we discuss the premise of ordinal data, and when and where and how such data come into being. To reiterate, ordinal data is a collection of ordered categories which are meant to denote levels of an attribute. For example, in the Social Sciences, it could be a level of agreement (or disagreement) with a statement, or it could be degree of liking for an object or construct (Really Like, Like, Neutral, etc.). In these contexts, one often uses ordinal variables in surveys to understand unobserved constructs like "parenting capability" or "happiness". On another hand, ordinal data could arise in life sciences as a stage of growth (bud, flower, fruit or infant, child, teen, adult). In the medical setting, ordered data is very common. For example, pain scale, tumor stage, disease stage, disease severity. As with any other data type, we need to understand the question that is of relevance to the researcher to analyze ordinal data. In this book, we will talk about the different contexts and the questions asked and talk about analysis for these different types of questions.

This chapter is by no means a comprehensive literature review, but a look at the history of the different methods that arose from the different questions associated with ordinal data. It is more of a philosophical look at ordinal data, the various questions that can be asked, the method of analysis and the nuances of each of the methods. In each case, we look at the seminal paper on the topic and talk about the thought process connected with the methods. The hope is putting these all in the same chapter will allow the readers to see the differences and similarities of the methods.

DOI: 10.1201/9781003020615-3

There are a few methods that are often used in the context of ordinal data. These methods stem from different perspectives about ordinal data and the question asked. Even though some of these methods are inappropriate for modeling and prediction, we describe these methods especially in the context of how they came about. Then, in the following three chapters, we discuss the models in detail and talk about their pros and cons.

We will focus our review on three main types of analysis for ordinal data **and include a fourth type to show why it is not relevant.** The 4 types we discuss are:

1. Likert Scale and analysis related to it.
2. Cumulative Distribution (CDF)-based analysis, (Logit and Probit models)
3. Latent Variable model (LVM)
4. Rank-Based models

While all four types of analyses are generally discussed in the framework of ordinal responses, this book focuses on modeling an ordinal response with a combination of categorical and numerical predictors. Using a combination of types of predictors creates more of a regression or prediction set up and as such, CDF-based models are the most appropriate.

We will also clarify why rank-based models are not appropriate for ordinal data, even though the ordinal responses are actually ranks. To understand the nuances of each of the methods, we will relate it to data examples as much as possible. If our data examples (horse data and heart data) are not relevant for the methods, we will use illustrative examples without actual data.

3.1 Likert Method

Ordinal data is inexplicably linked with Rensis Likert and his seminal work from 1932 [51]. An online search of the word "Likert" pronounced ("Leek"-art, not "Like"-art) came up with over 100 million hits and most were on constructing surveys and how to work with ordered responses. Even with having millions of resources available, the term "Likert Scale" is probably one of the most misunderstood and misrepresented method of analysis in statistics. In Likert's original paper [51], the focus was on finding a way to measure personality traits rather than the *data analysis method* to look at ordinal data. To quote, this study is:

> *... aimed first of all to present a wide array of problems having to do with these five major attitude areas: international relations, race relations, economic conflict, political conflict and religion. The attitude areas best covered in the questionnaire are those of race relations, international relations, and economic conflict.*

The article discusses the myriad of methods used during that time to measure "attitude" and mentions the work of several contributors with a focus on the work by Thurstone (1927) [80], (1928) [78] (a-c), (1929) [81], and (1930) [79]. In Likert's words, Thurstone's methods comprised of using methods from psychophysics to understand the difference in the steps in the attitude scale and equalize them.

In 1932, the common method used was based on a "scale" which was determined by a distribution of the underlying latent distribution. Hence, when a 5-point scale was used, the numbers on the 5-point scale were replaced by their corresponding σ values which was then summed for analysis. What is brilliant about the Likert (1932) [51] study was that it laid down a very strong foundation of the idea of an unseen, latent variable. The crux of the work is on understanding how to quantify and measure attitudes, not the technique used.

Likert's (1932) [51] study talks about a survey with questions on topics related to International Relationships, Race relation, and Imperialism. The questions were: *Yes, ?, No*, multiple choice questions, and several 5 point scales which were: "strongly disapprove" to "strongly approve". In his analysis, he noticed that several of the 5 point scale answers followed a somewhat "Normal" distribution. He used this Normal tendency to combine these types of mixed questions to measurement scales. Using this empirical evidence, he decided that the data were *Normal like* enough to assume Normal distribution. He then went to show that rather than using the σ values, if one used the numbers as they were 1 to 5, and summed those to construct the scales, the reliability of this latter (much simpler) method was as good as the one using the more complicated σ scores. Thus, **the use of the numbers themselves, became what is known as the Likert method**. However, we have to keep in mind that the Likert scale is a summation of ordinal data for multiple questions on a topic. **It does not deal with individual questions.** The idea was to understand the underlying latent construct rather than analyze this or predict.

The distinction between what we are interested in and Likert scale is basic. Our interest is an ordinal response (hence, one Likert-type item) with mainly numerical predictors (and some categorical predictors) while analyses using the Likert scale requires multiple ordinal responses on a topic. Hence, while Likert is often associated with ordinal data, this book will not focus on Likert Scale based analysis even though we will cover this topic briefly in chapter 4.

3.2 Cumulative Ordinal Models

Cumulative models like Logit, Probit, and Complementary log-log models are probably the most common models used in traditional statistics for ordinal regression. As most predictive modeling uses CDF-based models, this method of analysis is the main focus of this book.

Ordinal Probit and Logit

While logit models and logistic regression is far more common in CDF-based models, historically the probit preceded the logit. The word **probit** was coined by Bliss (1934)[11] by combining the word **prob**ability and un**it** in the context of converting percentages to probabilities and was then popularized by Finney (1947)[26]. Ordinal probit appeared in the literature when Aitchison and Silvey (1957) [4] proposed the ordered probit model to analyze experiments in which the responses of subjects to various doses of stimulus are divided into ranked classes. These models have been widely used as a methodological framework for analyzing ordered data since the pioneering work of McKelvey and Zavoina (1975) [56]. Though out-popularized by logit, the economics literature still tends to use probit and ordered probit (see Greene (2000) [32], DeCanio (1986) [20], Boex (2000) [38], Chan, Miller and Teha (2005) [18]).

Snell (1964) [72] suggested the use of the logistic distribution instead of the Normal distribution as an approximation for reasons of mathematical simplification. The first comprehensive treatment of ordered response models in the social sciences appeared with the work of McKelvey and Zavoina (1975) [56] who generalized the model of Aitchison and Silvey (1957) [4] to more than one independent variable. Their basic idea was to assume the existence of an underlying continuous latent variable—related to a single index of explanatory variables and an error term, and to obtain the observed categorical outcome by discretizing the real line into a finite number of intervals.

While McCullagh and Nelder are probably better known as being pioneers of the Generalized Linear Model. McCullagh's paper (1980) [54] originally developed the model in the context of ordinal data. He developed and discussed a general class of regression models for ordinal data based on a latent variable. The purpose of his paper was to investigate models appropriate to measurements on an ordinal scale. In particular, he introduced the most commonly used ordered logit model, the proportional odds model. The proportional odds model is widely useful in practice because its interpretation is simple and it has potentially greater power than multinomial logit models for ordered response variables.

The use of statistical methods for categorical data has increased dramatically, particularly for applications in the biomedical and social sciences. Agresti (2019) [2] has summarized these methods and has shown readers how to apply them using software. The major focus of this book is categorical predictors and contingency table type analysis.

Many statistical software include easy-to-use procedures for CDF regression models. Although it is relatively straightforward to specify and estimate (since the advent of computers) CDF models, the interpretation of the results is more complicated and less intuitive compared to linear regression. Detailed background of logistic regression can be seen in Kleinbaum et al. (1998) [47], Hosmer and Lemeshow (1980) [36], Agresti (1980) [2], McCullagh (1989) [54], and McCulloch and Searle (2001) [55]. We will talk about these models in detail in chapter 5.

3.3 Latent Variable Models (LVM)

While, both logit and probit models are developed with the idea that there is a latent variable in the background, these are considered CDF models, as we assume the form of the underlying CDF of the latent variable and model it as a function of observed predictors. However, there is a class of models that deals directly with understanding latent variables called Latent Variable models (LVM). LVM is divided into Structural Equation Modeling (SEM) or Item Response Theory (IRT). SEM is widely used in social sciences, psychometrics and education literature, though it is not accepted as widely by mainstream statistics. However, we do need to clarify at this juncture, that in LVM the underlying response variable(s) are not *necessarily* ordinal. As a matter of fact, LVM with ordinal response is fairly recent and most LVMs assume a continuous latent response. IRT will not be considered in this book as it really deals with testing theory and not directly connected with modeling ordinal data.

Structural Equation Model (SEM)

SEM focuses on the unseen latent variables and the joint analysis of latent and observed variables. Here, the response and explanatory variables could be a combination of latent and observed variables. Historically, SEM comprised of three different methods: linear regression analysis (LRA), Confirmatory Factor Analysis (CFA), and Path Analysis (PA). While these do **not** often deal with ordinal responses, we talk about these methods in this book. Our purpose for doing so is to explain why LVM are not often a suitable method for the questions we are interested in and to give an idea of where LVM would be suitable.

To look at the history of SEM, we need to consider the history of the three methods that comprise SEM: Linear Regression Analysis (LRA), Confirmatory Factor Analysis (CFA), and Path Analysis (PA). LRA is credited to Pearson (1938) [67] after the development of the correlation coefficient (1898) [68]. However, the concept of LRA was suggested by Galton (1894) [28] who clearly used the idea of the regression line to predict the response, Y, using explanatory variables x and the least-squares (LS) method. The LS method was published as early as 1805 by Adrien-Marie Legendre [50]. However, it is clear that Carl Friedrich Gauss used it as early as 1794 in this research. His publication on this topic is (1809) [29] was a precursor to this field. Hence, one can say that the idea of LRA preceded the formal formation of the field of statistics.

Factor Analysis (FA) was developed fairly soon after linear regression and is credited to Spearman (1904) [74], (1927) [75]. Spearman's idea was that if a set of items correlated they could be summed to yield a construct. Spearman was the first to use the term "factor analysis" in defining a two-factor construct for a theory of intelligence. Most of the aptitude, achievement, and diagnostic tests, surveys, and inventories still use versions of factor analysis. CFA is credited to Jöreskog (1969) [42]. However, it is based on earlier work by Howe (1955) [37], Anderson and Rubin (1956) [6], and Lawley (1958) [49]. Factor Analysis often called exploratory factor analysis explores the relationship among multiple correlated variables, in an effort to explore the underlying constructs. CFA, on the other hand, is meant to test the existence of hypothesized constructs from observed data.

The third component of SEM, Path Analysis (PA), is credited to Sewell Wright (1918) [87], (1921) [86], (1934) [88]. These models combine both correlation coefficients and regression analysis to model more complex relationships. This method was mostly ignored by mainstream statistics (as the original papers were related to animal behavior). In the 1950s, it was "discovered" by econometricians in the context of simultaneous regression. Essentially, PA involves solving a set of simultaneous regression equations that theoretically establish the relationship among the observed variables in the path model.

The idea of SEM is combining CFA with LRA and PA to understand complex relationships between both latent and observed variables. SEM is credited to Karl Jöreskog (1969) [42], (1973) [43], Ward Keesling (1972) [44], and David Wiley (1973) [85] and was called the **JKW** model. This was then called **LISREL** (after the advent of the software by that name). Now, there is multiple software and myriad of articles on its theory and application. Muthen published a slew of articles along with the software **Mplus** about applications of SEM (for example (1998–2012) [65]). While most of the papers in SEM involve a continuous response, **Mplus** does give options for ordinal responses. However, Muthen (2002) [63] has also argued about the lack of acceptance of

SEM in traditional statistics literature and show various examples in statistics where one uses latent variables.

3.4 Rank-Based Methods

While it is tempting to use Rank-Based methods for ordinal data, it is important to understand that there is a fundamental difference between rank-based data and ordinal data. In a rank-based method, all the units are ranked among themselves, whereas in ordinal data each individual is given a rank between 1 and r. To go back to our horse example, we would use a rank-based method if we ranked the 1071 horses among themselves by performance and then potentially compared if the horses with low Consensus EIPH performed better than horses with severe Consensus EIPH. In this scenario, we would use the Kruskal Wallis test or Mann Whitney test to compare the groups on basis of their ranks.

Currently, there is a push for "ranked regression", but by replacing the continuous response by its rank among the other observations. These methods are robust alternatives for linear regression, but not viable for analyzing ordinal data. Hence, we do not include a chapter on this topic.

3.5 Applications of the Different Methods

Imagine a hypothetical grocery store that is interested in understanding customer liking of the store. They have a survey that comprises of:
On a scale from 1 to 5, with 5 being the most agreement (i.e., (1) Strongly Disagree, (2) Disagree, (3) Neutral/Not Sure, (4) Agree, (5) Strongly Agree), answer the following questions (1–8) and provide demographic information:

1. It was easy to find the items you wanted in the store (Y_1)
2. Checking out your items was simple (Y_2)
3. The items purchased were of good quality (Y_3)
4. There was enough space in each aisle to move around (Y_4)
5. The shelves were stacked neatly (Y_5)
6. There was enough parking for the store (Y_6)
7. The personnel were helpful and courteous (Y_7)
8. I will come back to this store (Y_8)

Other relevant information:

9. Average number of times we do groceries per week is (X_1):
10. The gender of the person who comes to the store more often is (X_2):
11. Our family size is (X_3):
12. Our family has a child under 2 (X_4): Yes/No
13. Average monthly income of our family is (X_5):

Using this example we will show the distinction between the methods used.

Likert Scale Analysis

Each of items 1–8 in the survey is considered a Likert item. To construct a Likert Scale, we would sum up the eight Likert items and find the scale. In general, that was all Likert scales were intended to be used for. However, people often use the Likert Scaled value as a Normal response and could do a model relating the Likert Scale to the relevant information as explanatory variables. So, the Likert Scale is calculated using:

$$Y_{lik} = \sum Y_i \text{ with i= 1,..,8}$$

CDF: Logit and Probit Analysis

This type of survey data is not really suited to a CDF analysis. But, to show how it would work for comparison, we would take ONE of the Likert items, say item 8 (I will come back to this store, Y_8). This would be our ordinal response and we would model the probability of $Y_8 = j, j = 1,..,5$ with the items 9–13 on the relevant list as explanatory variables, $X = (X_1, X_2, X_3, X_4, X_5)$.

$$P(Y_8 \leq j|X) = Function(j, X)$$

The form of the function would depend upon whether the underlying CDF is assumed Logit or Probit.

SEM

This data is best suited for SEM. As one could take the 8 questions and hypothesize that these relate to latent constructs around convenience (Y_1, Y_2, Y_6), Pleasantness of the store (Y_4, Y_5, Y_7), and quality (Y_3, Y_8). One would then use CFA to build the latent constructs, look at the interrelationships between the response items and the explanatory variables and do a path analysis. This method would require using CFA to move from the ordinal response to a continuous factor loading as the response. However, software like **MPlus** would allow for the ordinal response to be used directly using CDF models.

3.6 Summary

In this chapter we give a brief overview of the different methods that are associated with the analysis of ordinal data. The convenience store example above is meant to allow the readers to see the difference of the questions of interest that gave rise to the specific methods. The point is that even though the methods are connected by their shared utility of analyzing ordinal data, they answer different questions and are thus not interchangeable. For example, we use some principles of SEM in CDF models and vice versa. But they are meant to answer a very different set of questions.

This historical review is a just the tip of the iceberg. For each of the papers we cited, there are probably hundred more we did not cite. The purpose of this chapter is to highlight that people have used ordinal data in very different ways and often the different fields with different perspectives and assumptions that have not had much overlap. There has been misconceptions around what the methods do and lack of clarity around the limitations of the methods. In this book we will focus on modeling and focus on the "how" and "why" with worked examples.

4

Likert Scale

4.1 Introduction

In chapter 3, we talked about Rensis Likert's seminal paper from 1932 [51]. In this short chapter we talk about the type of analysis commonly associated with Likert's method. We distinguish between the terms item and scale and talk about when and where it is appropriate to use this kind of method and the caveats therein.

Before Likert as we saw in chapter 3, the method used in analyzing surveys that had ordered responses, was to convert the numbers, 1 to r, to their σ values and **then** sum the σ values for analysis. Likert's contribution was to show that if one used the numbers as they were 1 to r, and summed those to construct the scales, the reliability of this latter (much simpler) method was as good as the one using the more complicated σ scores. Thus, **the use of the numbers themselves became what is known as the Likert method.** However, before we talk about analysis we have to keep in mind that the Likert scale is a summation of ordinal data for multiple questions on a topic, **not an individual response**.

4.2 Likert Items versus Likert Scale

Let us first dispel some common myths by distinguishing between a Likert item and a Likert Scale. The term Likert, whether item or scale is most often associated with ordered responses from surveys and questionnaires. As a matter of fact, Likert scale is often used interchangeably with rating scales, even though there are many other types of rating scales. In common usage, a Likert **item** is an item or question on a survey which allows the respondent to select an answer along a range. We would like to point out that Likert categorically distinguished between a scale and an item. That is, a scale emerges from collective responses to a set of items (usually eight or more). Derrick and White (2017) [21] and Carifio and Perla (2007) [17] talk extensively about the "urban myths" surrounding Likert scale and its abuses.

DOI: 10.1201/9781003020615-4

Common examples of Likert style items are found in surveys where respondents choose answers from symmetric agree-disagree scales for a series of statements. These items are often given 3, 5, 7, or 10 choices and the respondent chooses one of the choices for each statement. Odd numbers represent the inclusion of the "neutral category". Other variations are also appropriate, including the deletion of the neutral response (Clason and Dormody (1994) [19]) where we have an even number of values available to the response, forcing the response to show some leaning—positive or negative. The idea behind the number of response options for a given item depend upon the item being measured and subjects' abilities to discriminate between response options. While, even numbered responses are possible and are often used when it is a stage (example: infant, toddler, tween, teen, as stages for children under 18), we illustrate the more commonly used 5-point scale in this chapter.

Let us consider the example from chapter 3 where we are interested in measuring attitudes about a new store.

On a scale from 1 to 5, with 5 being the most agreement (i.e., (1) Strongly Disagree, (2) Disagree, (3) Neutral/Not Sure, (4) Agree, (5) Strongly Agree), respond to the following statements:

1. It was easy to find the items you wanted in the store
2. Checking out your items was simple
3. The items purchased were of good quality
4. There was enough space in each aisle to move around
5. The shelves were stacked neatly
6. There was enough parking for the store
7. The personnel were helpful and courteous
8. I will come back to this store

Each statement and its response on the rating scale is considered to be a Likert **item** (Clason and Dormody (1994) [19]). A Likert **scale** would then be calculated by summing or averaging groups of items (DeVellis (1991) [22], Stewart and Ware (1992) [76]) and then the Likert scales would be used for future analysis. Groups of Likert items to be summed or averaged are chosen based on the common attitudes or ideas that they measure. The basic assumption of Likert scaling is that distances between each choice (answer option) are equal. The idea is that with summation or averaging, the properties of central limit theorem can be invoked and one can assume Normal distributions. Researchers, routinely, use items that are highly correlated among themselves (high internal consistency). However, the idea is together the items will capture the question under study.

There are common rules for constructing the items: (1) items are presented as strong declarative and (2) responses to items include a range of values.

While constructing the Likert scale, Dukes (2005) [23] surmised each item to be equally weighted. If an investigation involves k such items, each measured on the same r-point Likert response format (responses coded as 1, 2, ..., r, with higher scores reflecting more endorsement of each item), then the theoretic range of the Likert scale is k to kr. Assumptions underlying the construction of multiple-item Likert scales are that each item is linearly related to the overall scale score, and that each item comprising the scale has approximately the same distribution (e.g., similar means and standard deviations). The recent popularity of qualitative research techniques has relieved some of the burden associated with the dilemma of not meeting these assumptions; however, many social scientists still rely on quantitative measures of attitudes, using Likert scale as a numerical measure. The original Likert scale is comprised of a series of questions with five response alternatives: strongly approve (1), approve (2), undecided (3), disapprove (4), and strongly disapprove (5). Likert combined the responses from the series of questions to create an attitudinal measurement scale (on international relations, race relations, economic conflict, political conflict, and religion). His data analysis was based on the composite score from the series of questions that represented the attitudinal scale. It did **not** analyze individual questions.

While alternatives of the Likert response have become common in ordinal data research, common usage has also created misuses or mistakes. One mistake commonly made is the improper analysis of individual questions on an attitudinal scale. That is, analyzing a Likert item instead of a Likert scale. To properly analyze Likert data, one must understand the measurement scale represented by each item. Numbers assigned to Likert-type items express a "greater than" relationship, however, how much greater is not defined. Descriptive statistics recommended for ordinal measurement scale items include a mode or median for central tendency and frequencies for variability. Additional analysis procedures appropriate for ordinal scale items include the chi-square measure of association, Kendall Tau B and Kendall Tau C as discussed by Kendall (1938) [45].

4.3 Data Analysis

Generally, in a properly constructed survey with Likert items that enable us to construct Likert scales, the average of the responses of the Likert items is used as the response variable. Other control variables are measured as the predictors and then methods used for continuous/interval data like linear models, Analysis of Variance (ANOVA) [13] are performed.

In this book, our interest is an ordinal response (hence, one Likert-type item) with numerical and categorical predictors. Likert scale deals with multiple ordinal Likert items on a topic. Hence, while Likert is often associated with ordinal data, this book will not focus on Likert Scale-based analysis. However, we will illustrate how Likert style analysis could be used.

Horse Dataset example

We analyze the Horse Dataset using Likert Scale analysis. In this dataset, the response variable is the EIPH score for the horses after the race.

To talk about Likert scale in the context of the Horse Dataset, we would have to measure EIPH by using several, k, different methods on a scale of 1 to r. We would then average the scores over the k items and use the averaged score to create a Likert Scale item. Historically, this Likert Scale item would be analyzed as a numerical response with standard general linear models applied to the analysis.

While our response is not appropriate for Likert scale because the EIPH is not measured in several different ways, we **do** have three veterinarians scoring each of the horses. With some stretch of the imagination, we could consider each score as a Likert item and then the average score would be our Likert scale. We can then perform linear regression on with the averaged scores as the response variable. We summarize our results here, with the caveat that **this is for illustration only** and not the correct way to analyze this dataset.

In the dataset, the scores for the three veterinarians are "EIPH Gold", "EIPH Leguillette", and "EIPH Sellon". For the Likert Scale, we average these scores. Then, we perform a multiple linear regression to use Lasix (if a horse had this procedure done—yes/no), Surface (the type of surface the horses ran on - all weather/turf/dirt), Finish (the finish position of the horse) and Finish Time (time taken to complete the race) to predict the the average EIPH score.

The R Code and results are below:

```
1  library(dplyr) # Loading the required library
2
3  # Changing variable types and calculating the average
       EIPH
4
5  DATA_Likert = DATA %>%
6    mutate(`EIPH Gold` = as.numeric(`EIPH Gold`),
7           `EIPH Leguillette` = as.numeric(`EIPH
     Leguillette`),
8           `EIPH Sellon` = as.numeric(`EIPH Sellon`),
9           AVG_EIPH = (`EIPH Gold`+ `EIPH Leguillette`+
10                        `EIPH Sellon`)/3)
11
```

```
# Multiple Linear Regression
model_fit_Likert = lm(AVG_EIPH ~ Lasix + Surface +
    Finish + 'Finish Time', data=DATA_Likert)
summary(model_fit_Likert)
```

```
Call:
lm(formula = AVG_EIPH ~ Lasix + Surface + Finish + `Finish Time`,
    data = DATA_Likert)

Residuals:
    Min      1Q  Median      3Q     Max
-1.2947 -0.7439 -0.0805  0.5212  3.3018

Coefficients:
               Estimate Std. Error t value Pr(>|t|)
(Intercept)    0.125247   0.258140   0.485  0.62766
LasixY        -0.083241   0.095826  -0.869  0.38527
SurfaceDirt    0.529990   0.199510   2.656  0.00804 **
SurfaceTurf    0.541582   0.210253   2.576  0.01016 *
Finish         0.023430   0.011028   2.125  0.03390 *
`Finish Time`  0.003194   0.001926   1.659  0.09758 .
---
Signif. codes:  0 '***' 0.001 '**' 0.01 '*' 0.05 '.' 0.1 ' ' 1

Residual standard error: 0.8761 on 866 degrees of freedom
  (196 observations deleted due to missingness)
Multiple R-squared:  0.02296,	Adjusted R-squared:  0.01732
F-statistic:  4.07 on 5 and 866 DF,  p-value: 0.001171
```

R Output

FIGURE 4.1
Likert analysis results

The generic SAS code is given below:

```
/* Likert Analysis*/
DATA likert ;

/* CALCULATING THE MEAN Y*/
SET dataset ;
meany =( y1+y2+y3)/3;
RUN ;

/* MULTIPLE LINEAR REGRESSION */
PROC REG DATA = likert ;
MODEL meany =x1 x2 x3 x4;
 run ;
```

The results show that surface, finish place, and finish time are statistically significant at at $\alpha = 0.10$ level. The estimates show that a horse running on turf surface has a higher average EIPH than a horse running on all weather surface. Also, a horse running on dirt surface has a higher average EIPH than a horse running on all weather surface. Finally, as finish and finish time increases, the average EIPH is also increasing.

The details involved in making these interpretations will not be discussed here as Likert Scale is not the main focus of this book.

While in asking multiple questions we can indirectly get to the attitude we want to measure, conducting analysis this way with this type of data is not without issues. It is agreed upon that when we use a scale of 1–5, 1, 2, 3, 4, 5 are not numbers. These are categories, where 5 has more agreement than 4, which has more agreement than 3, which has more agreement than 2, and so on. One can argue that it is fair to treat them as numbers, since indeed 5 is bigger than 4, which is bigger than 3, and so on. However:

- The numbers are just relative scales and a 5 for one person could be the same as a 4 for another. It depends upon their rating, and these numbers are only comparable across the same person.
- What does the difference between the ratings mean? We do not know if the difference between the ratings have the same size . For example, the assumed distance between 1 and 2, could be the same or different than the distance between 3 and 4.

4.4 Summary

In this chapter, we talked about the issues around Likert item, Likert scale, and the myths and misuses therein. The take home message is, one cannot use an individual Likert item as a continuous or interval variable and analyze data using methods appropriate for continuous data. If multiple Likert items are available and the summed or averaged score can be assumed to be Normal, one could use methods like t-test, LRA, and ANOVA. But much care needs to be taken in this type of analysis. Finally, for the type of predictive analysis we are interested in, Likert method (item or scale) is not an appropriate method of analysis. Likert analysis focuses on coming up with the underlying latent variable underlying ordinal survey questions.

5

Cumulative Distribution Function (CDF) Models

5.1 Introduction

In chapter 3, we talked about the idea of and the history behind CDF models. Here, we will first talk about linear models in general and then talk about the type of models used in Ordinal Regression. To do so, we will take an overall look at General Linear Models (GLMs) and Generalized Linear Models and the problems based on the models and types of data. While, most statisticians and data scientists are aware of these ideas, we wanted to present this for completion here. Let us set up some notation first:

Notations

- $(\mathbf{X})$ denote our matrix of explanatory variables.
  - $(\mathbf{X})$ is a n by p matrix.
  - $(1, x_{(1,i)}, ..., x_{(p-1,i)})$ denotes the ith observation for the $(p-1)$ explanatory variables including intercept.
  - The $(p-1)$ of $(\mathbf{X})$ can be numerical (continuous or discrete), ordinal, or categorical.
- (Y) denote the stochastic response variable.
  - CASE 1: (Y) is numerical, normal-like random variable
  - CASE 2: (Y) is a discrete random variable
  - CASE 3: (Y) is a categorical random variable
  - CASE 4: (Y) is an ordinal random variable
- β denote the parameter vectors including intercept and slope, respectively
- Let ϵ represent the random error vector

DOI: 10.1201/9781003020615-5

5.2 Models

5.2.1 General Linear Model

CASE 1: The model in this case is called the General Linear Model (GLM) and is given by:

$$(Y) = \mathbf{X}\beta + \epsilon.$$
$$\epsilon \longrightarrow N(0, \sigma I)$$

Depending upon the structure of the components of (X) we have:

- Analysis of Variance (ANOVA): All items in X are categorical
- Regression: All items in X are numerical
- Analysis of Covariance (ANCOVA) or multiple linear regression with dummy variables: X has a mix of numerical and categorical components.

This is the traditional or classical GLM and used widely.

5.2.2 Generalized Linear Model

CASES 2, 3, 4: Without the normality assumption of the errors, the additive error model cannot be used. This is because the location-scale properties of the Normal distribution is not true for other distributions. However, for distributions in the exponential family (Poisson, Binomial, etc) we can extend the idea of the General Linear Model (GLM) to the Generalized Linear Model. This model allows us to model the expected value of a random variable from the exponential family of distributions as a function of $X\beta$. We write this as:

$$E(Y) = F(\mathbf{X}\beta)$$

Here, F is the link function. In the Logistic Regression case this is the CDF of the logistic distribution.

This book is about the Case 4. Hence, let us take an in depth look at the idea of CDF models.

5.2.3 The CDF Models

These are models that are used in classical statistics to model an ordinal response as a function of multiple explanatory variables. The explanatory variables could be numerical, categorical, and ordinal. The most common CDF models are the Logit and Probit models we described in chapter 3. As we explained earlier, here the assumption is that there is a **single** unobserved latent continuous variable that is explained by our predictors. We assume that this unobserved latent variable is divided into groups (or chunks or sections) and the observed ordinal categories are manifestations of these groups. Hence, to understand the relationship between the explanatory and the latent variable, we model the probability of being in a group as a linear function of explanatory variables. As probabilities are between 0 and 1, the CDF function is a handy tool for modeling.

As mentioned in chapter 3, McCullagh's paper (1980) [54] originally developed the idea of the Generalized Linear Model (GLiM or GENMOD) in the context of ordinal data. He developed and discussed a general class of regression models for ordinal data based on a single latent variable. Following the trend of the literature of that time, the purpose of his paper was to investigate structural models appropriate to measurements on an ordinal scale. Further, he introduced the proportional odds model. The proportional odds model is widely useful in practice because its interpretation is simpler and it has potentially greater power than multinomial logit models for ordered response variables. These are models that are used for investigating the relationship between the ordinal response and the covariates (see McCullagh (1980) [54] and Kim (1995) [46]). The objective is to link up the manifestation or latent variable and the covariates.

Regressing and predicting an ordinal response is the focus of this book and hence, this chapter will be the main theoretical content of this book and will have the most mathematical details.

5.3 Theory of CDF Models

5.3.1 CDF as Threshold Models

To really understand CDF models, we would like to introduce the idea of thresholds and tolerance. Though, ordinal data is more "number-like" than binary data, we will start with the idea of binary data.

To motivate this we will revert to our Horse dataset which was discussed in chapters 1 and 2. Let us consider the unobserved variable *the horses' stress to its lung capillaries.* This stress can be brought on by a multitude of factors, which are the explanatory factors in our data, speed at which it ran, the position, the weather, the type of surface, etc. and the combination of this produces stress to the lung capillary. Beyond a certain level of stress, the capillaries burst and bleeding happens. While, we cannot observe the latent stress variable that is responsible for the bleeding, we can observe in each horse whether there is no bleeding, 0 or some bleeding, 1. The amount of stress beyond which bleeding occurs is unseen, but this is considered our "tolerance" level. Hence, binary regression models the **probability** of bleeding to the explanatory variables in an effort to understand what are the predictors that lead to going beyond the tolerance level.

Let us consider the unknown latent variable, the stress put on the lungs to be η. We denote our binary response by, B, and B can take the value 0 (no bleeding) or 1 (bleeding). Let the unknown threshold or tolerance be denoted by τ.

We model the unknown η as a function of the $x_1, ..., x_p$ as:

$$\eta = \beta_0 + \beta_1 x_1 + ... + \beta_p x_p + \epsilon$$

Hence,

$$B = \begin{cases} 0, & \text{if } \eta \leq \tau \\ 1, & \text{if } \eta > \tau \end{cases}$$

The idea is to model the unseen, η as a function of the x. But as η is not observed, we model the probability of the observed B. As this probability is often in terms of "less than" probabilities, modeling these with CDF of distributions became a common choice. So that:

$$P(B = 0) = P(\eta \leq \tau | X) = P(\beta_0 + ... + \beta_p x_p + \epsilon \leq \tau) = P(\epsilon \leq \tau - \beta_0 + ... + \beta_p x_p)$$

The distribution assumed for ϵ determines the type of CDF used. If ϵ is assumed to be from a logistic distribution we have the binary logit model and if ϵ is assumed to be from a Normal distribution, then we assume a Probit model.

5.3.2 Logistic Distribution

The standard Normal distribution assumption associated with the Probit model is well-known and found in any introductory texts in statistics. However, the logistic distribution associated with the logistic regression models are less well-known. In this section, we will briefly talk about this distribution and explain why the logit transform is a natural parametrization for this model.

The logistic distribution is a continuous probability distribution that is symmetric. The distribution looks like the Normal distribution in shape but has a heavier tailed than the Normal distribution (another point in its favor for application). The logistic distribution is a special case of the Tukey lambda class of distributions. Like the Normal distribution it has two natural parameters, μ (location) and κ (scale). We purposely stay away from calling the scale σ because the scale is not the standard deviation of the logistic distribution like it would be for the Normal distribution .

The probability density function of the Logistic Distribution for μ, κ is given by:

$$f(x, \mu, \kappa) = \frac{e^{-((x-\mu)/\kappa)}}{\kappa(1 + e^{-((x-\mu)/\kappa)2}} \tag{5.1}$$

The probability density function of the Logistic Distribution, for μ=0 and κ = 1, is given by:

$$f(x, 0, 1) = \frac{e^{-x}}{(1 + e^{-x})^2} \tag{5.2}$$

With the CDF:

$$F(x, 0, 1) = \frac{e^{-x}}{(1 + e^{-x})}$$

Hence, if a probability, p, is equated to the CDF then,

$$p = \frac{e^{-x}}{(1 + e^{-x})} \tag{5.3}$$

$$\frac{p}{1-p} = e^{-x} \tag{5.4}$$

$$log(\frac{p}{1-p}) = -x \tag{5.5}$$

Hence, the we see that the $log(\frac{p}{1-p})$ is a natural parametrization of the logistic distribution. We shall discuss this in more detail in the later sections.

Ordinal Data Perspective

To understand the ordinal perspective, we need to assume that there are multiple tolerance levels and beyond each tolerance level we observe a response. Going back to our horse example, let us assume that the stress on the lungs variable, η has thresholds, $\tau_1, \tau_2, \tau_3, \tau_4$ which allows Y our ordinal response to take values 0, 1, 2, 3, or 4. In other words, based on the tolerance to the stress, we observe no bleeding, 0, to lungs full of blood, 4.

Then, the ordinal response categories can be defined using these "tolerance levels" or threshold values such that:

$$Y = \begin{cases} 0, & \text{if } \eta \leq \tau_1 \\ 1, & \text{if } \tau_1 < \eta \leq \tau_2 \\ 2, & \text{if } \tau_2 < \eta \leq \tau_3 \\ 3, & \text{if } \tau_3 < \eta \leq \tau_4 \\ 4, & \text{if } \eta \geq \tau_4 \end{cases} \tag{5.6}$$

As before, we link η to x using a linear equation and the assumption on the ϵ determines the CDF model used.

Results from the CDF models are useful in answering questions such as:

- Based on some explanatory variables, how likely is it to be in one category of the response compared to another category of the response or compared to a combination of categories of the response?
- How important are the explanatory variables when predicting the ordinal response?
- What are the sizes and directions of the effect of the explanatory variables on the estimation of the ordinal response?
- How well can we predict the likelihood of being in specific categories of the ordinal response by using the explanatory variables?

5.3.3 Ordinal Logistic Regression

The premise of the ordinal logit stems from the assumption that the distribution of ϵ, which is linked to the latent η, follows a logistic distribution that was discussed earlier. Hence, like the binary logistic regression model, the ordinal logit model uses log-odds ratios to model the relationship between the explanatory variables and the response variable, as this is a natural parametrization with logistic distribution explained earlier.

For the ordinal logit model, relationships are estimated using log odds which are called logit or log-odds ratio. Let us now use mathematics to understand why this makes sense. $P(Y \leq j|X)$ is the cumulative probability of observing a response in category j or below. The term *logit* specifically refers to the function that takes the natural logarithm of this probability divided by its complement (i.e., the probability of that event not happening). Essentially, it takes the natural logarithm of the "odds" of being in at most category j.

Because cumulative probabilities are used in this formulation of the log-odds, this logit is called the cumulative logit. Mathematically, the cumulative logit for the ordinal response is defined as:

$$logit[P(Y \leq j|X)] = log\frac{P(Y \leq j|X)}{1 - P(Y \leq j|X)}, j = 1 \ldots, c-1$$

where Y is the ordinal response variable with c categories, and X is the matrix of the explanatory variables.

Using logit, we describe how likely it is for an observation to fall in at most category j given that the explanatory variables, X, were observed. The model describes this logit as being a linear function of the x in the logistic distribution set up.

Because the utility of the log-odds or logit comes from the use of odds, it is important to understand the odds. The odds is defined as the likelihood of an event happenning compared to the event not happenning In general, that is:

$$ODDS_A = \frac{p}{1-p}$$

where p is a probability of the event A occurring.

So, it can be seen that when $ODDS = 1$, it is equally likely that the event will happen or will not happen (i.e., there is a $50/50$ chance of that event happening). Then, $0 < ODDS < 1$ occurs when the chance of the event happening is less than the chance of the event not happening and $ODDS > 1$ occurs when the chance of the event happening is greater than the chance of the event not happening.

So, when interpreting the logit or log-odds, we are interpreting how likely it is for the event (related to the possible outcomes of the response variable) to occur as compared to the event not occurring. Odds are common in the gambling parlance and often horse-racing (as our primary example) bets are provided as odds. So, if a favorite is given odds of 4:3, it means the odds ratio is $4/3$, or the horse is more likely to win than to lose. The probability of the horse winning in that case is $4/7$ and losing is $3/7$.

When interpreting the logit and the results of the CDF model, it is also important to understand that we are assuming the linear relationship between the explanatory variables and the latent variable. The CDF models assume that there is an underlying continuous latent variable and the tolerance level or the "cut-points" that are also unobserved. The latent variable is a linear combination of some explanatory variables and an error. If the model has one explanatory variable x, then the latent variable η is defined as:

$$\eta = \beta_0 + \beta_1 x + \epsilon \tag{5.7}$$

where η is the underlying latent variable and ϵ is the error that follows a logistic distribution. For example, in the Horse Race Dataset example when we are looking at Consensus EIPH with 5 responses (0–4), the latent variable would be a continuous variable, stress response, and the τ are the tolerance levels that produce the ordinal responses. It would be divided in such a way that it is proportional and in the same order as the ordinal response categories. So, this continuous latent variable would be cut up into 5 regions using 4 "cut-points". In general, the ordinal response categories can be defined using these "cut-points" or threshold values such that:

$$Y = \begin{cases} 1, & \text{if } \eta \leq \tau_1 \\ 2, & \text{if } \tau_1 < \eta \leq \tau_2 \\ \vdots & \vdots \\ c, & \text{if } \eta \geq \tau_{c-1} \end{cases}$$

So, the ordinal logit model is defined by:

$$logit[P(Y \leq j|X)] = \alpha_j - \beta^T X, \qquad j = 1, \ldots, c-1 \tag{5.8}$$

where the predictor/effect parameters β determine how the cumulative logit changes when there are changes in the explanatory variables, X, and the intercept parameters α_j determine the threshold cumulative logit for the response categories (i.e., the "cut-points" for the response categories).

5.3.4 Understanding the Parameters Mathematically

To understand how ordinal logistic regression works, we need to look at parameters of the ordinal regression in equation (5.8) β and α_j.

To keep things simple let us consider the situation when we have one explanatory variable, x for an ordinal response, Y with a latent construct η. As the generalization to multiple responses is straightforward.

In this case,

$$\eta = \beta_0 + \beta_1 x + \epsilon$$

Further,

$$\begin{aligned} P(Y \leq j) &= P(\eta \leq \tau_j | X) \\ &= P(\beta_0 + x\beta_1 + \epsilon \leq \tau_j) \\ &= P(\epsilon \leq \tau_j - \beta_0 + x\beta_1) \\ &= P(\epsilon \leq \alpha_j - x\beta_1) \end{aligned} \tag{5.9}$$

This allows us to see that:

- The intercept parameter $\alpha_j = \tau_j - \beta_0$, i.e., the intercept parameter of the latent parameter and the tolerance τ_j.
- Software will often make assumptions like β_0=0 for identifiability and then do the regression in terms of τ_j
- It is easiest to define α_j and admit that in this formulation, separate estimation of β_0 and τ_j is not possible.

If we assume ϵ follows a logistic distribution then,

$$P(\epsilon \leq \alpha_j - x\beta_1) = \frac{e^{-(\alpha_j - x\beta_1)}}{(1 + e^{-(\alpha_j - x\beta_1)})} \tag{5.10}$$

- Taking log odds would allow us to have a linear function of $(\alpha_j - x\beta_1)$.

The parameters β and α_j are estimated using maximum likelihood estimation. The estimates for β measure the changes in the predicted cumulative logits based on the values of the explanatory variables. In this formulation, if the estimate for β is negative, it implies that as the value of the explanatory variable increases, there is a tendency for the category of response to increase. However, different software use different conventions regarding the sign of the β. When we discuss the data example, we will talk about this issue again. Before we talk about estimation, we do need to clarify another assumption made in ordinal logistic regression.

5.3.5 Proportional Odds Assumption

The term proportional odds is used because this model assumes that the effect of all the explanatory variables are the same on the estimates of the intercepts. In other words, all the input variables have an equal effect on a specific outcome in the response. This is referred to as the parallel lines assumption or proportional odds assumption because it ensures that the slope of the logistic function is the same across all the category thresholds for the response variable. This makes estimating the coefficients easier, since it reduces the number of coefficients of the model to a single set of coefficients across all the categories of the response variables.

Figure 5.1 shows an example of how the proportional odds assumption is represented in the estimates of the coefficients. The coefficients are estimated so that the regression functions for the different categories of the response are parallel on the logit scale.

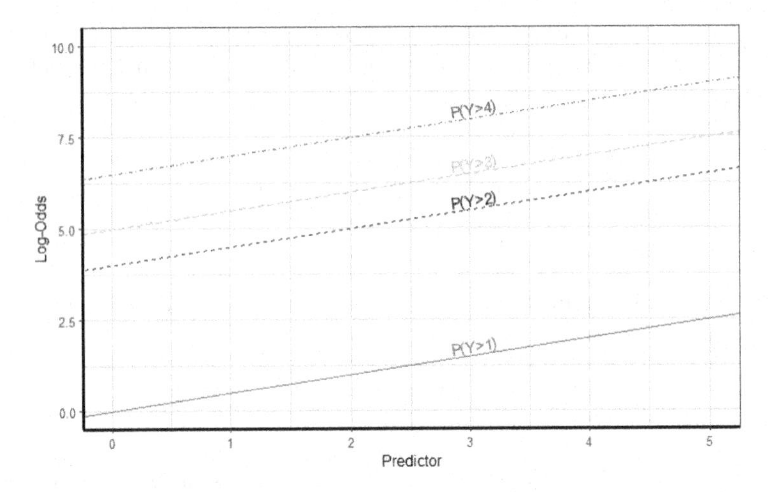

Example of the estimated slopes and intercepts when the proportional odds assumption is assumed. Each line corresponds to different cut-points as described. All the lines are parallel to each other (i.e., have the same slope) due to the proportional odds assumption.

FIGURE 5.1
Log odds ratio estimates for ordered logistic regression.

In software like R and SAS, The estimates of the coefficients are log-odds ratios are calculated by estimating differences in logits, ζ_i. In general, assuming that X_1 and X_2 are two realizations of the random variable X, this calculation is:

$$\zeta_i = logit[P(Y \leq j|X = X_1)] - logit[P(Y \leq j|X = X_2)] \tag{5.11}$$

$$= \alpha_j - \beta^T X_1 - (\alpha_j - \beta^T X_2) \tag{5.12}$$

$$= \beta^T (X_1 - X_2) \tag{5.13}$$

We find that ζ_i is proportional to the distance between $X = X_1$ and $X = X_2$ and this property applies for all $i = c - 1$ logits in the model. This implies that:

$$\zeta_i = \beta^T(X_1 - X_2) = log\frac{P(Y \leq j|X = X_1)}{P(Y > j|X = X_1)} - log\frac{P(Y \leq j|X = X_2)}{P(Y > j|X = X_2)}$$

$$= log\frac{\frac{P(Y \leq j|X=X_1)}{P(Y>j|X=X_1)}}{\frac{P(Y \leq j|X=X_2)}{P(Y>j|X=X_2)}}$$

$$= log\frac{\frac{\text{Probability that } y \leq j \text{ given } X=X_1}{\text{Probability that } y>j \text{ given } X=X_1}}{\frac{\text{Probability that } y \leq j \text{ given } X=X_2}{\text{Probability that } y>j \text{ given } X=X_2}}$$

$$= log\frac{\text{Odds for } Y \leq j \text{ given } X = X_1}{\text{Odds for } Y \leq j \text{ given } X = X_2}$$

$$= log(\text{odds ratio for } Y \leq j \text{ given } X = X_1 \text{ compared to when } X = X_2)$$

Practically, this allows us to understand how to interpret the estimates of the coefficients. Since ζ_i simplifies to the Log-odds ratio, we can interpret it by translating our understanding of odds ratio to what it implies when the odds ratio is transformed into the log-scale. In general:

- odds ratio < 1 implies that the event in the numerator$(X = X_1)$ is related to a lower odds of the response variable occurring as compared to the event in the denominator$(X = X_2)$
- odds ratio > 1 implies that the event $X = X_1$ is related to a higher odds of the response variable occurring
- odds ratio $= 1$ implies that the events $X = X_1$ and $X = X_2$ have the same odds of the response variable occurring

Using the functional relationship, log-odds $=$ log(odds ratio) or equivalently $e^{\text{log-odds}} =$ odds ratio, the log-odds ratio that are estimated by R can then be interpreted similarly.

- odds ratio < 1 implies log-odds < 0
- odds ratio > 1 implies log-odds > 0
- odds ratio $= 1$ implies log-odds $= 0$

5.3.6 Estimating the Parameters: Maximum Likelihood Estimation

The parameters for each of these methods are determined through the process of Maximum Likelihood Estimation (MLE). To understand how MLE operates, we will explore the log-likelihood function of the proportional odds model. The same idea follows when other types of logits are used.

Maximum Likelihood Estimation for Proportional Odds Model

As there is no direct distribution for an ordinal response variable, we dichotomize the response variable and use the Binomial distribution to model the probabilities of responses being or not being in a given category. Suppose there are n observations and we have an ordered response with c categories from each of the observations. For observation i, let us define, ${Y^*}_{ij}$ = the binary indicator of the jth response for observation i. Therefore, for category j,

$$ {Y^*}_{ij} = \begin{cases} Y_{ik} = 1, k = j \\ Y_{ik} = 0, k \neq j \end{cases} $$

For each observation i, given the predictor variable x_i, the probability that the observation responded for the category i is given by,

$$P(Y_i = j|X = x_i) = \Pi_j(x_i),\; j = 1, 2, .., c\;; i = 1, 2, ..., n$$

For ease of notation we define the the probability of being in group j given x_i as $\Pi_j(x_i)$.

$$\Pi_j(x_i) = (P(Y_i \leq j|x_i) - (P(Y_i \leq j-1|x_i)) \tag{5.14}$$

Assuming that all the observations are independent (an important assumption), the likelihood function for all the n observations can be expressed as the product of n multinomial probability mass functions. To note, in the MLE we use the binary classification probability and the multinomial distribution. The Multinomial distribution assumes that all the observations are categorical, not ordinal. However, as the probabilities of each category are calculated based on the difference between the adjacent groups, it allows the ordinal nature of the data to be incorporated.

$$\begin{aligned} L(\{\alpha_j\}, \beta) &= \prod_{i=1}^{n}\prod_{j=1}^{c} \Pi_j(x_i)^{Y_{ij}} \\ &= \prod_{i=1}^{n}\left(\prod_{j=1}^{c} (P(Y_i \leq j|x_i) - (P(Y_i \leq j-1|x_i))^{Y_{ij}}\right) \\ &= \prod_{i=1}^{n}\left(\prod_{j=1}^{c}\left(\frac{e^{\alpha_j+\beta' x_i}}{1+e^{\alpha_j+\beta' x_i}} - \frac{e^{\alpha_{j-1}+\beta' x_i}}{1+e^{\alpha_{j-1}+\beta' x_i}}\right)^{Y_{ij}}\right) \end{aligned} \tag{5.15}$$

The log likelihood function in 5.15 is given by,

$$\begin{aligned} l(\{\alpha_j\}, \beta) &= log(L) \\ &= \sum_{i=1}^{n}\sum_{i=1}^{c} y_{ij} log\left(\frac{e^{\alpha_j+\beta' x_i}}{1+e^{\alpha_j+\beta' x_i}} - \frac{e^{\alpha_{j-1}+\beta' x_i}}{1+e^{\alpha_{j-1}+\beta' x_i}}\right) \end{aligned} \tag{5.16}$$

To find the maximum likelihood estimates of the parameters, we must have,

$$\begin{cases} \frac{\partial}{\partial \alpha_j}(l(\{\alpha_j\}, \beta)) = 0, j = 1, 2, .., c \\ \frac{\partial}{\partial \beta_k}(l(\{\alpha_j\}, \beta k)) = 0, \text{for an effect parameter} \beta_k \end{cases}$$

Upon simplification, we get,

$$\sum_{i=1}^{n}\sum_{j=1}^{c} y_{ij}\left(\frac{g(\alpha_j + \beta' x_i)}{G(\alpha_j + \beta' x_i) - G(\alpha_{j-1} + \beta' x_i)}\right) = 0 \tag{5.17}$$

$$\sum_{i=1}^{n}\sum_{j=1}^{c} y_{ij} x_{ik}\left(\frac{g(\alpha_j + \beta' x_i) - g(\alpha_{j-1} + \beta' x_i)}{G(\alpha_j + \beta' x_i) - G(\alpha_{j-1} + \beta' x_i)}\right) = 0, \tag{5.18}$$

where, $\begin{cases} G(z) = \frac{e^z}{1+e^z} \\ g(z) = \frac{e^z}{(1+e^z)^2} \end{cases}$

The equations presented above (5.17 and 5.18) lack a closed-form solution. McCullagh (1980) [54], introduced an iterative re-weighted least squares algorithm to address this. McCullagh (1980) [54] highlighted that unique estimates might not exist or could be infinite for a finite n. Just as in binary logistic regression, the absence of an estimate or its infiniteness can result from quasi-complete separation. Similarly, for the cumulative logit model, this scenario arises when there are no pairs of observations that are either all concordant or all discordant in terms of category ordering.

5.3.7 Inference: Confidence Intervals and Testing

Most software give maximum likelihood estimates and standard errors for the estimates. Testing and confidence intervals are done using asymptotic properties of MLE. The assumptions allow us to use the Normal distribution assumption for the maximum likelihood estimates and proceed with constructing confidence intervals and hypothesis tests. As the tests are on the β parameters which are interpreted in the context of log odds ratios, it is more common to use confidence intervals. Software often provides confidence intervals for both log odds ratios and the odds ratios.

Confidence intervals refers how confident we are that our constructed intervals are the ones that capture the true value of the parameter. Below, we provide the formula for calculating the 95% two-sided and one-sided confidence intervals. However, the formulas can be used for any size confidence interval by updating the choice of quantile.

The 95% two-sided confidence interval for a given β_i is calculated by:

$$\begin{aligned}
\text{CI} &= (\text{Log-odds ratio}_{Lower}, \text{Log-OddsRatio}_{Upper}) \\
&= \text{LOR Estimate} \pm (1.96 \times \text{standard error}) \\
&= \hat{\beta}_i \pm 1.96 \times \sigma_{\hat{\beta}_i}
\end{aligned}$$

where 1.96 is the 97.5^{th} quantile from the Normal distribution. This gives the $(1-\alpha) \times 100$ confidence interval when $\alpha = 0.05$. Since it is a two-sided confidence interval, α is divided by 2 when calculating the required quantile.

The 95% one-sided confidence interval for a given β_i is calculated by:

$$\begin{aligned}
\text{CI} &= (\text{Log-odds ratio}_{Lower}, \infty) \\
&\text{Log-odds ratio}_{Lower} \\
&= \text{LOR Estimate} - (1.645 \times \text{standard error}) \\
&= \hat{\beta}_i - 1.645 \times \sigma_{\hat{\beta}_i}
\end{aligned}$$

OR

$$\begin{aligned}\text{CI} &= (-\infty, \text{Log-OddsRatio}_{Upper})\\ \text{Log-OddsRatio}_{Upper} &\\ &= \text{LOR Estimate} + (1.645 \times \text{standard error})\\ &= \hat{\beta}_i + 1.645 \times \sigma_{\hat{\beta}_i}\end{aligned}$$

where 1.645 is the 95th quantile from the Normal distribution. This gives the $(1-\alpha) \times 100$ confidence interval when $\alpha = 0.05$

For interpretability, the confidence intervals for odds ratios are usually reported. We can calculate this by exponentiating the Log-odds ratio.

$$\text{CI} = (OR_{Upper}, OR_{Lower}) = e^{\hat{\beta}_i \pm 1.96 \times \sigma_{\hat{\beta}_i}}$$

Because of the nature of the function e^x, this confidence interval is not symmetric around the estimate of the odds ratio like the confidence interval for the Log-odds ratio is.

5.3.8 Types of Logits

The proportional odds models utilize the cumulative logit and throughout this book, we use the cumulative logits to to illustrate our examples. Though this is the most commonly used logit, there are other ways to form logits that can recognize the category ordered nature of ordinal data. When logits other than cumulative logit are used, the resulting model and its interpretation will differ from the cumulative logit version we describe in this chapter. Below, we describe the cumulative logit and other logits that can be used in place of the cumulative logit.

Cumulative Logit

This logit compares the cumulative probability of all events till a specified event compared to the cumulative probability of all other events.

$$\begin{aligned}logit[P(Y \le j)] &= log\frac{P(Y \le j)}{1 - P(Y \le j)}\\ &= log\frac{P(Y \le j)}{P(Y > j)}, j = 1, ..., c-1\end{aligned}$$

This cumulative logit is the logit used in the model described in this chapter. An example of this logit in use is comparing the likelihood of being at having pain at most degree 6 to the likelihood of having pain at least degree 7.

Adjacent-Categories Logit

This logit models the log-odds for pairs of adjacent categories.

$$logit = log\frac{P(Y = j)}{P(Y = j + 1)}, j = 1, ..., c - 1$$

This logit allows the researcher to compare the likelihood of being in a category compared to the immediate next (adjacent) category. An example of this is the likelihood of being happy compared to being very happy.

Continuation-Ratio Logit

This logit is suitable when there's a sequential relationship among the response categories. It models the log-odds of continuing to endorse category j given that the response is at least j $(Y \geq j)$.

$$logit = log\frac{P(Y = j)}{P(Y \geq j + 1)}, j = 1, ..., c - 1$$

This logit allows a researcher to compare the likelihood of being in a category versus all the other categories beyond that category. This logit is especially useful when the events are set up in such a way that one thing has to happen before the others can happen. For example, you can compare the likelihood of living to a certain time given survival up to that time.

Baseline Logits

This logit is commonly used to model nominal categorical data where the researcher is able to compare the likelihood of being in a category versus a category that is set to be the baseline. Examples of this is the likelihood of being happy as compared to very sad and the likelihood of being very happy to being very sad.

This baseline logit is equivalent to the sum of adjacent categories logits as follows:

$$logit = log\frac{P(Y = j)}{P(Y = c)}, j = 1, ..., c - 1$$

Where c is the baseline category.

The use of any of these logits would depend on the comparisons you are wanting to make with your results. For the most part, calculations can be done to move from one logit to another, but it makes sense to start out with the logit of interest first.

5.3.9 Ordinal Probit Regression

Like the ordinal logit model, the ordinal probit model is a regression model that uses explanatory variables to predict the ordinal response. The ordinal probit model uses estimates that are based on probabilities of the Normal distribution instead of the logistic model. The functional form of the ordinal probit model is:

$$\Phi^{-1}[P(Y \leq j|X)] = \alpha_j - \beta^T X, \qquad j = 1, \ldots, c-1$$

where Φ is the cdf of the Normal distribution, Y is the ordinal response and X are the explanatory variables. Since the ordinal probit model uses the cdf of the Normal distribution, the error for the latent variable, ϵ, is assumed to be normally distributed.

The choice of a Normal link function instead of the logistic function is usually made when there are unequal error (heteroskedastic) variances. This is commonly found in fields such as advanced econometrics and political science. However, simulation studies ([59]), show that the results from the ordinal logit and ordinal probit models are usually similar due to the similarity of the Normal and Logistic distributions. Because of the similarity of the two models and the fact that the properties of Normal distribution are well known, we do not discuss this model further.

5.4 Data Analysis

In this section, we specify an ordered logit model to predict the "Consensus EIPH" level from the Horse Dataset described in chapters 1 and 3. Consensus EIPH is ordinal (from 0 to 4). The objective is to use Lasix Treatment, Type of Surface, Finish Position, and Finish Time to predict the level of Consensus EIPH. In this case, Consensus EIPH is the response variable, Y, while Lasix Treatment, Type of Surface, Finish Position, and Finish Time are explanatory variables, X. We chose these variables as they represent each type of variable: binary, categorical with more than 2 categories, ordinal, and numerical. There were many other relevant explanatory variables that we did not use in this illustration.

Below is the R Code and Output or the ordinal logit model. The default logit used is the cumulative logit. However, specifying the method to be logistic also ensures that an ordinal logistic model is fit.

```
1 library(MASS)    ## Loading the MASS library
2
3 model_fit = polr(ordered(ConsensusEIPH) ~ Lasix +
    Surface + Finish + 'Finish Time' , data = DATA,
    Hess = TRUE, method = c("logistic"))
4 summary(model_fit)
```

```
Call:
polr(formula = ordered(ConsensusEIPH) ~ Lasix
+ Surface + Finish +
    `Finish Time`, data = DATA, Hess = TRUE,
method = c("logistic"))

Coefficients:
                    Value Std. Error t value
LasixY          -0.075796   0.185382 -0.4089
SurfaceDirt      1.441398   0.464736  3.1015
SurfaceTurf      1.387886   0.480683  2.8873
Finish           0.075514   0.020677  3.6520
`Finish Time`    0.009427   0.003686  2.5573

Intercepts:
    Value   Std. Error t value
0|1  1.8671  0.5629      3.3170
1|2  3.3702  0.5691      5.9225
2|3  5.0040  0.5790      8.6427
3|4  6.8380  0.6255     10.9322

Residual Deviance: 2767.339
AIC: 2785.339
```

R output for fitting an ordinal logistic regression using the polr function.

FIGURE 5.2
These are the ordinal logistic regression model estimates that were generated from R's polr function. The results include the call for the function, the estimates for the coefficients, their standard error and corresponding t-values, the estimates for the intercepts and their corresponding standard error and t values. Additionally, the Residual deviance and AIC for the model are also included.

The generic SAS code for fitting an ordinal logistic model is given below. The dataset is called "dataset", the ordinal response is Y, and the explanatory variables are X1, X2, X3 and X4.

```
/* ORDINAL LOGISTIC RUN*/
PROC LOGISTIC DATA=dataset PLOTS=ALL;
MODEL Y=X1 X2 X3 X4 / PCORR  CLODDS=both  CTABLE;
run;
```

With this generic code SAS will give us:

- The Analysis of Maximum Likelihood Estimates of the intercepts and slopes as well the p-values.

- The 95% (it is the default and can be changed) confidence interval for the odds ratio.
- Some generic association statistics and model fit statistics.
- It will provide the relevant prediction plots
- Provide the table of observed and expected counts
- Will provide confidence interval for odds ratio

Interpreting the Intercepts of the Model

We first look at interpreting the intercepts. These are interpreted as the estimates for the cutpoints or tresholds of the latent variable.

We use the results from Figure 5.2 to illustrate how to interpret the intercepts for the model using the horse data. In this case, the latent variable is "stress" and the cutpoints determine the 5 categories for Consensus EIPH, the response variable. From the results, these cutpoints are 1.8671, 3.3702, 5.0040 and 6.8380. Notice that there are $c - 1 = 4$ cutpoints, since there are $c = 5$ categories of the response variable (i.e. 0,1,2,3,4,5). These 4 cutpoints correspond to the splits of the latent variable where the split between 0 and 1 is made at 1.8671, between 1 and 2 is made at 3.3702, between 2 and 3 is made at 5.0040, and between 3 and 4 is made at 6.8380.

The cutpoint $0|1$ is interpreted as the log-odds of low Consensus EIPH. This corresponds to the odds that is related to $\frac{P(ConsensusEIPH \leq 0)}{P(ConsensusEIPH \geq 1)}$. In other words, this represents the log odds of a horse having a Consensus EIPH of 0 versus at least 1. The cutpoint $1|2$ is interpreted as the log-odds of a horse having Consensus EIPH of at most 1 versus having at least 2. The other cutpoints are interpreted similarly.

The estimates of the cutpoints are not usually interpreted as a part of the reporting of the results as they are not particularly meaningful. However, it can be used to see how the underlying latent variable was cut up. Because we assume that there is a continuum of the response variable that is represented by the latent variable, these estimates represent the cutpoints used to dichotomize the latent variable into the response variable.

Interpreting the Coefficients of Explanatory Variables

Next, we can get the estimates of the coefficients for the explanatory variables. Depending on the type of variable, we interpret them differently.

General Interpretations

Before we illustrate with the horse data example, a general understanding of the way the coefficients are estimated is warranted.

First, an understanding of how baseline groups of explanatory variables is determined in software is necessary. This is because the estimates of the coefficients are interpreted as differences between the baseline group and the other non-baseline groups. In R, the baseline group is the group that is first alphabetically or numerically by default (and other software may have different defaults). For example, if the categorical explanatory variable is color with groups red, yellow and blue, then the baseline group is blue. As a result, the estimates of the coefficients will refer to differences between (1) red and blue, and (2) yellow and blue.

Another important detail is the ordering of the response variable. If an order is not set, R defaults the ordering to be done alphabetically or numerically. For example, a response variable with groups high, medium, and low will have a default ordering of high, low, and medium, respectively with medium being the highest level and high being the lowest level. So, the phrase "higher level of the response variable" means the likelihood of observing a response near the end of the ordered list (e.g., medium as compared to high or low).

Then, if a categorical explanatory variable is involved, it can be seen from equation 5.19 below that interpreting the log-odds involves interpreting the relationships between the baseline and non-baseline groups of both the response variable and the explanatory categorical variable. If a numerical explanatory variable is involved, then only the baseline and non-baseline groups of the response variable will be considered. This is because the interpretation will be based on a unit increase of the explanatory variable.

$$\text{log-odds} = log(\frac{\text{Odds}(P(Y \leq j|x = \text{a non-baseline group}))}{\text{Odds}(P(Y \leq j|x = \text{the baseline group}))} \tag{5.19}$$

For a single explanatory variable, the actual estimate that we get from R is a simplified version of equations 5.13 and 5.19 above. The baseline group is coded as $x = 0$ and the non-baseline group is coded as $x = 1$, which resolves the log-odds described in equations 5.19 and 5.13 to:

$$\begin{aligned}\text{log-odds} = \eta_i &= logit[P(Y \leq j|X = 1)] - logit[P(Y \leq j|X = 0)] \\ &= \alpha_j - \beta_i - (\alpha_j) \\ &= -\beta_i\end{aligned}$$

So, to account for the negative estimate of β_i, the log-odds will be interpreted in relation to a higher level of the response variable instead of the opposite interpretation as would be expected. Then the general interpretations are:

Categorical Explanatory Variable

1. Positive log-odds estimates imply:
   (a) the odds related to a higher level of the response is more likely for the non-baseline group than the baseline group, or equivalently

   (b) the odds related to a lower level of the response is more likely for the baseline group than the non-baseline group.

2. The opposite is true for a negative log-odds. Negative log-odds imply:
   (a) the odds related to a lower level of the response is more likely for the non-baseline group as compared to the baseline group, or equivalently
   (b) the odds related to a higher level of the response is more likely for the baseline group as compared to the non-baseline group
3. When comparing the log-odds of multiple groups within an explanatory variable, a group with larger log-odds is more likely to be associated with higher values of the response variable than the baseline group.

Numerical Explanatory Variable

1. Positive log-odds imply:
   (a) as the explanatory variable increases by 1 unit, the odds related to a lower level of the response is less likely, or equivalently
   (b) as the explanatory variable increases by 1 unit, the odds related to a higher level of the response is more likely
2. The opposite is true for a negative log-odds. Negative log-odds imply:
   (a) as the explanatory variable increases by 1 unit, the odds related to a lower level of the response is more likely, or equivalently
   (b) as the explanatory variable increases by 1 unit, the odds related to a higher level of the response is less likely

Interpretations for the Horse Data Results

- **Categorical variables** like surface and Lasix are interpreted by comparing the groups within the variable in terms of their likelihood of experiencing a greater response assuming that all other explanatory variables are held constant. In this case, higher response corresponds to higher Consensus EIPH scores. Practically, greater response corresponds to the horse experiencing more bleeding in the lungs. The groups of the categorical variables are compared to their baseline group. We determine the baseline group by identifying the group that is first alphabetically (or alternatively, the group that does not appear in the R output).

  1. In this case, **surface** dirt is compared to the baseline **surface** all weather. The estimate of 1.441398 is interpreted as dirt surface is associated with a higher odds of a horse having a higher

Consensus EIPH as compared to all weather surface. This means that the horse is more likely to have bleeding in their lungs if they run on dirt versus all weather surface.

2. Likewise, the **surface** turf is compared to the baseline surface all weather. The estimate of 1.387886 is interpreted as a turf surface is associated with a horse having a higher Consensus EIPH compared to an all weather surface. This means that the horse is more likely to have bleeding in their lungs if they run on turf surface versus all weather surface
3. In addition, since the estimate for **surface** dirt is greater than the estimate for surface turf, then dirt surface is associated with a higher Consensus EIPH than turf surface. This means that it is more likely for a horse to have bleeding in their lungs when running on dirt as compared to turf.
4. In the case of **Lasix**, yes is compared to the baseline **Lasix** no. The estimate for this coefficient is negative which suggests that it is less likely for a horse with Lasix treatment to be associated with a higher Consensus EIPH than a horse with Lasix Treatment no. So, practically, this means that horses are less likely to have bleeding in their lungs if they had the Lasix treatment.

- **Continuous Numeric Variables** are interpreted comparing the likelihood of having a higher response based on unit increases of the predictor variables assuming all other explanatory variables are held constant. There is no need to identify a baseline group for numerical explanatory variables.

  1. The estimate for **finish time** is 0.009427. This is positive and implies that as the predictor variable is increasing by 1 unit, the log-odds associated with higher Consensus EIPH is increasing. So, as finish time increases by one unit, it is more likely that a horse will have a higher Consensus EIPH. Practically, as finish time increases by one unit, it is more likely that a horse will experience bleeding in the lungs.

- **Discrete Numeric Variables and Ordinal Variables** are interpreted similarly to Continuous Numeric Variables.

  1. The estimate for **finish** is 0.075514, which is positive. It suggests that as finish position increases, the odds that a horse will have a higher Consensus EIPH increases. Practically, as finish increases by one unit, it is more likely that a horse will experience bleeding in the lungs.

Interpreting p-values and confidence intervals

To decide if an estimate is statistically significant, we can use p-values or confidence intervals. Determining statistical significance is a method used to quantify if the estimate that was calculated is likely or due to chance. Traditionally, an α of 0.05 is used to check for statistical significance. However, depending on the nature of your study, you may decide to make the α value smaller or larger. For example, a study related to the social interactions of humans might use an α that is much larger allowing for more chance while a scientific study related to the measurement of PFAS (Per- and Polyfluorinated Substances) in rain water might use a much smaller α allowing for less chance.

The p-value and the confidence intervals are calculated using the statistical t-distribution (as seen in the output in Figure 5.2). They consider the tails of the distribution and uses the chosen α to determine where on the tails of the distribution constitutes observing a likely estimate versus an estimate by chance.

- When using **p-values**, statistical significance is achieved when the p-value associated with the estimate is smaller than the chosen α.
- When the **confidence interval** is used, statistical significance is achieved when log-odds $= 0$ is not included in the interval.

For determining a statistical significance using the confidence interval, the value 0 is used because log-odds $= 0$ corresponds odds ratio $= 1$. This implies that there is no difference between the baseline group and the non-baseline group in terms of the response variable. If the confidence interval includes 0, it implies that it is likely that the difference (i.e., the estimate) you observed occurred just by chance. If the confidence interval does not include 0, then it implies that the estimate you observed is unlikely to have happened just by chance. The interpretation is similar when a p-value is used, except that the the comparison is made using α rather than log-odds.

Confidence intervals can also be used in the traditional sense to talk about how confident we are in the estimates of the log-odds ratios. A confidence interval in statistics refers to an interval where the true value of the population parameter is likely to be. So, we are able to say how confident we are in the estimates calculated from our data.

Unlike SAS (which provides p-values and confidence intervals) R does not calculate p-values and confidence intervals by default. For the Horse Dataset example, the R Code and results are below.

The R Code is given below:

```
1 ## Saving the R output
2 coefficient_table = coef(summary(model_fit))
3
4 ## Calculating p-values
```

```
5 pvalues = pnorm(abs(coefficient_table[, "t value"]),
    lower.tail = FALSE) * 2
6
7 ## Saving the p-values and the confidence intervals
    to the original R output with log- odds estimates
8 coefficient_table = cbind(coefficient_table, "p value
    " = pvalues)
9 coefficient_table
10
11 ## Calculating the confidence interval for the log-
    odds estimates
12 Confidence_Interval = confint(model_fit)
13 Confidence_Interval
14
15
16 #To get confidence interval assuming Normality, use
    this
17 Confidence_interval_Normality = confint.default(model
    _fit)
```

	Value	Std. Error	t value	p value
LasixY	-0.075795836	0.185382142	-0.4088627	6.826405e-01
SurfaceDirt	1.441397563	0.464735795	3.1015419	1.925156e-03
SurfaceTurf	1.387886322	0.480682525	2.8873243	3.885336e-03
Finish	0.075514286	0.020677324	3.6520338	2.601717e-04
\`Finish Time\`	0.009426722	0.003686164	2.5573259	1.054803e-02
0|1	1.867089093	0.562876522	3.3170492	9.097361e-04
1|2	3.370204498	0.569050760	5.9225024	3.170791e-09
2|3	5.003955860	0.578978047	8.6427385	5.488143e-18
3|4	6.837970026	0.625488647	10.9322049	8.085947e-28

(a) p-values attached to the R Output

	2.5 %	97.5 %
LasixY	-0.44073826	0.28666619
SurfaceDirt	0.57238303	2.41991315
SurfaceTurf	0.48489518	2.39329586
Finish	0.03505011	0.11613509
\`Finish Time\`	0.00221018	0.01665904

(b) Confidence intervals of log-odds

FIGURE 5.3
P-values and confidence intervals of ordinal logistic regression estimates

SAS gives these under PROC LOGISTIC, so no special code is needed.

With an α of 0.05, the estimates that are statistically significant correspond to surface (dirt and turf), finish, and finish time. Lasix is not statistically significant because the p-value is 0.6826405 which is larger than 0.05 and the confidence interval (−0.44073826 0.28666619) includes 0.

In terms of confidence, we could say that we are 95 % (i.e., $100\times(1-\alpha)\%$) confident that the log-odds ratio for finish time lies in the interval from (0.00221018, 0.1665904). Other confidence intervals for the other explanatory variables can be interpreted similarly.

Calculating and Interpreting odds ratios

It is very common to see odds ratios being reported instead of log-odds because interpreting the natural logarithm is more difficult to conceptualize. For one, the natural logarithm is a non-linear function where a unit increase in the log-odds represents an increase multiplied by e $\approx$ 2.71. Whereas, the odds ratio does not use the natural logarithm removing the added complexity. To calculate the odds ratios and their confidence intervals from from the log-odds estimates, we use:

$$\text{odds ratio} = e^{\text{log-odds}}$$

Recalling that the odds ratio is defined as:

$$\text{odds ratio} = \frac{\text{Odds for } Y \leq j \text{ given } X = X_1}{\text{Odds for } Y \leq j \text{ given } X = X_0}$$

Where $X = X_0$ is the baseline group for that explanatory variable.

For the Horse Dataset example, the odds ratios care calculated in R as follows:

```
#Calculating odds ratio
OR = exp(coef(model_fit))

#Calculating  confidence interval for odds ratio
ConfidenceInterval = exp(confint(model_fit))

#Putting odds ratio and confidence interval in the
    same table
cbind(OR, ConfidenceInterval)
```

```
                    OR       2.5 %    97.5 %
LasixY         0.9270054 0.6435611  1.331980
SurfaceDirt    4.2265986 1.7724859 11.244883
SurfaceTurf    4.0063729 1.6240048 10.949523
Finish         1.0784386 1.0356716  1.123148
`Finish Time`  1.0094713 1.0022126  1.016799
```

odds ratios and their confidence intervals

FIGURE 5.4
Odds ratios and confidence intervals estimates from ordinal logistic regression

SAS gives these under PROC LOGISTIC, so no special code is needed.

Like with log-odds, we can also use the confidence intervals to determine statistical significance or to talk about our confidence in the estimates. In this case, an odds ratio is not statistically significant if the confidence interval contains 1. In terms of statistical significance, Lasix is not statistically significant because the interval (0.6435611, 1.331980) contains one. All other explanatory variables are statistically significant. This is the same as the results when the estimates based on the log-odds were interpreted. One example of a confidence interval is (1.7724859, 11.244883) for surface dirt. We can say that we are 95% (i.e., $100 \times (1 - \alpha)\%$) confident that the odds ratio for surface dirt when comparing to surface all weather lies in the interval from 1.7724859 to 11.244883.

We can also interpret the odds ratios like we did for the log-odds. In doing so, we are now able to quantify how much more likely the events are. For our Horse Dataset example, that would be:

- Categorical variables

    1. For **surface** dirt, the odds of a horse being more likely to be associated a higher Consensus EIPH when running on dirt surface is 4.2265986 times the odds for a horse on an all weather turf.
    2. For **surface** turf, the odds of a horse being more likely to be associated a higher Consensus EIPH on turf surface is 4.0063729 times the odds for a horse on an all weather turf.
    3. For **Lasix** yes, the odds of a horse with Lasix Treatment yes being more likely to be associated a higher Consensus EIPH is 0.9270054 times the odds for a horse on an all weather turf.

- Continuous Numeric Variables

    1. As finish time increases by one unit, the odds that a horse is more likely to be associated with higher Consensus EIPH 1.0094713 times more.

- Discrete Numeric Variables and Ordinal Variables

    1. As finish position increases by one unit, the odds that a horse will be associated with higher Consensus EIPH is 1.0784386 times more.

Making Predictions

We are able to use our model to make predictions. These predictions are informed by our data and the model we used. While predictions for individual new horses can be made very easily by only using the predict function as in the R code chunks below or by referring to the provided SAS output, we illustrate visualizing the prediction curves based on our model. That way, we can make predictions for any horse by referencing their values of the explanatory variables on our prediction curves. To do this, we need to create a new dataset that consists of data with values of the explanatory variables within the ranges of the original dataset. By making predictions for many values within range of our original dataset, we are able to visualize the predicted probabilities for any possible datapoint.

For this Horse Dataset example, we create a dataset that has observations in the range of our explanatory variables (surface, Lasix, finish, and Fiinish-Time). We generate 10080 rows of data so that there are many combinations of data for the different groups of Lasix and surface with all the values of finish and a range values for finish time. Note that, the number of rows of dataset that you create will be unique to your problem. This number depends on the number of combinations of options you have in your own original dataset. For the horse dataset example, there are 2 options for Lasix, 3 options for surface, 12 options for finish and we generated 140 different finish times within the range of 50 to 120. This gives a total of 2*3*12*140 = 10080 rows. We then make predictions on this dataset.

The R Code to create the dataset is below:

```
1 library(tidyverse)
2 # creating a dataset with observations to make
    predictions for.
3 #The dataset is created usind sequences
4 # The datset is called DATA_NEW
5
6 DATA_NEW <- tibble(
7   Lasix = rep(c("N","Y"), 10080/2),
8   Surface = rep(c("Turf", "Dirt", "All Weather"),
    each = 10080/3),
9   Finish = rep(seq(from = 1, to = 12), times = 10080/
    12),
```

```
10   'Finish Time' = rep(seq(from = 50, to = 120, length
      .out=140 ), times = 72))
```

This creates a tibble (or data frame) that looks like:

A tibble: 10,080 x 4

Lasix <chr>	Surface <chr>	Finish <int>	Finish Time <dbl>
N	Turf	1	50.00000
Y	Turf	2	50.50360
N	Turf	3	51.00719
Y	Turf	4	51.51079
N	Turf	5	52.01439
Y	Turf	6	52.51799
N	Turf	7	53.02158
Y	Turf	8	53.52518
N	Turf	9	54.02878
Y	Turf	10	54.53237

1-10 of 10,080 rows   Previous 1 2 3 4 5 6 ... 100 Next

Dataset with observations in the range of the explanatory variables

FIGURE 5.5
Dataset created for predictions

The next step is to calculate the predicted probabilities for these observations based on the estimates from fitting the model. The R Code for calculating predicted probabilities is below:

```
1
2  # Making the predictions
3  # Using the previously fitted model called model_fit
4  # Using the new dataset called DATA_NEW
5  # Specifying that predicted probabilities should be
     calculated
6
7  predictions = predict(model_fit, DATA_NEW, type = "
     probs")
8
9  # Binding the predictions to the table with the
     observations
10 DATA_NEW <- cbind(DATA_NEW, predictions)
11
12 # show first few rows of the table with the
     corresponding predicted probabilities based on the
      fitted model
13 head(DATA_NEW)
```

The result is the following table:

Description: df [6 x 5]

	0	1	2	3	4
	<dbl>	<dbl>	<dbl>	<dbl>	<dbl>
1	0.4830945	0.3246573	0.1478532	0.03702709	0.007367857
2	0.4819795	0.3250779	0.1483578	0.03718419	0.007400589
3	0.4432035	0.3383845	0.1666866	0.04308566	0.008639700
4	0.4421018	0.3387230	0.1672304	0.04326681	0.008678033
5	0.4040318	0.3489207	0.1868576	0.05006107	0.010128849
6	0.4029569	0.3491639	0.1874361	0.05026931	0.010173722

6 rows

First 6 rows of dataset with predicted probabilities

FIGURE 5.6
Dataset with predicted probabilities based on model

It is easiest to use visualizations to interpret these predicted probabilities, hence the prediction curve. Explanatory variables can be visualized individually with the predicted probabilities to see the relationship between the response and the individual explanatory variable. We can also look at combinations of explanatory variables with the predicted probabilities to see the interactions between the variables. Here are a few examples:

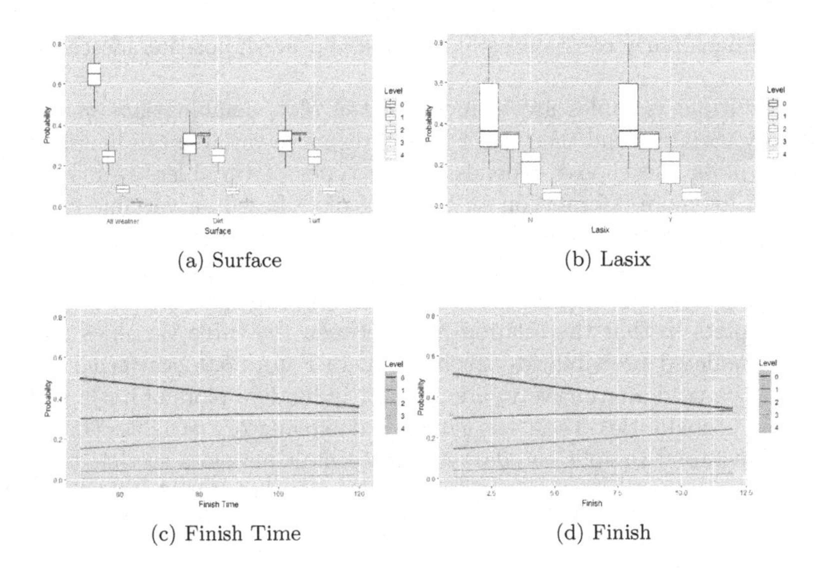

(a) Surface (b) Lasix

(c) Finish Time (d) Finish

FIGURE 5.7
Prediction curves for the individual explanatory variables and their predicted probabilities

Surface and Lasix are categorical variables and the predicted probabilities are numerical, so boxplots are appropriate. The locations of the boxplots in Figure 5.7 (a) show similar predicted probabilities for the different groups of Consensus EIPH when the horse is running on dirt surface and turf surface. On an all weather surface, the horse had higher predicted probabilities for Consensus EIPH 0 as compared to the horses running on dirt surface or turf surface. This is seen because the boxplot for Consensus EIPH 0 is higher on the y axis for all weather than for dirt surface and turf surface. Likewise, We see that the predicted probabilities for Consensus EIPH 1 is higher when a horse is running on dirt and turf than when a horse is running on an all weather surface.

The boxplots in Figure 5.7 (b) suggests that there is not a difference in predicted probabilities for all Consensus EIPH's for horses who have had Lasix procedure (Y) as compared to horses who did not have the procedure (Lasix:N).

Finish and finish time are numerical variables and the predicted probabilities is numerical, so line plots are appropriate. Figures 5.7 (c) and (d) have similar trends. When looking at the line corresponding to Consensus EIPH 0, it suggests that as finish time and finish position increases the predicted probabilities that a horse will have Consensus EIPH 0 decreases. On the other hand, predicted probabilities for Consensus EIPH 1, 2, and 3 increases as finish time and finish position increases. Predicted probabilities for Consensus EIPH remains fairly constant as finish time and finish position increases.

When multiple variables are included in the plot, combinations of the explanatory variables and their effect on the response variable can be seen. The types of plots you choose depends on the types of variables used. Refer to chapter 2 for more information on choosing plots based on variable types. In figure 5.8, line plots are used for the individual plots with finish time and the predicted probabilities because they are both numeric. Then the categorical explanatory variables (Lasix and surface) are used to split the data to create multiple plots so that the relationships between the three variables and the related predicted probabilities can be seen. In Figure 5.9, scatterplots (boxplots can be used alternatively) are used for individual plots of finish and the predicted probabilities. Then, the categorical explanatory variables (Lasix and surface) are used to split the data to create multiple plots so that the relationships between the three variables and their predicted probabilities can be seen.

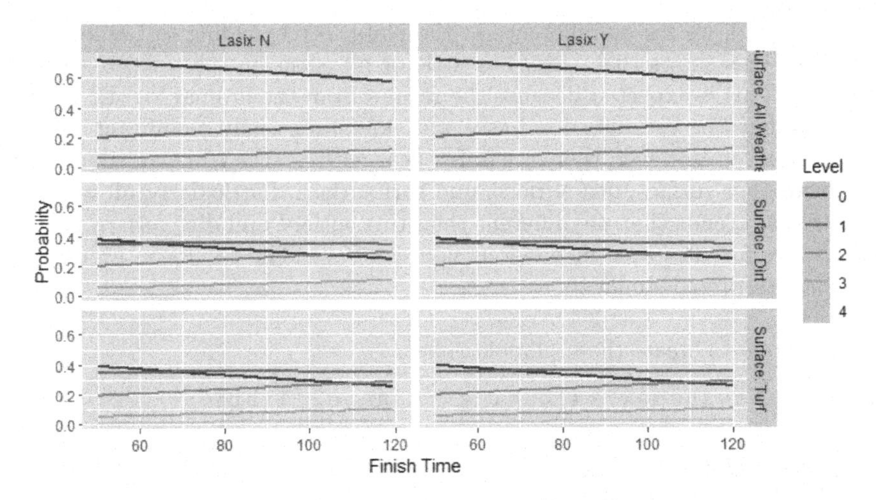

Surface, Lasix and Finish Time

FIGURE 5.8
Prediction curves for multiple explanatory variables (one numeric continuous, two categorical) and their predicted probabilities

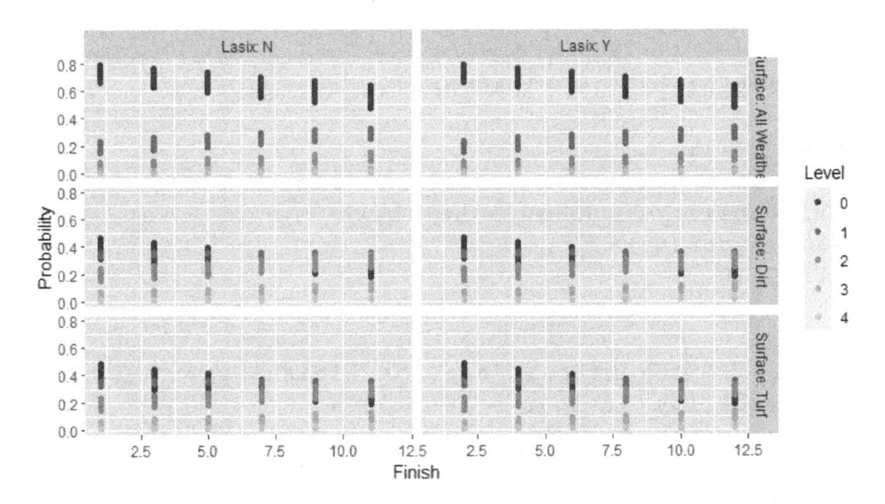

Surface, Lasix and Finish

FIGURE 5.9
Prediction curves for multiple explanatory variables (one numeric discrete, two categorical) and their predicted probabilities

In Figure 5.9, we see no difference in the plots corresponding to Lasix:N and Lasix:Y. This shows that Lasix has little or no effect on the predicted probabilities. This is expected because the results from the ordinal logistic model showed that the estimate for Lasix was not statistically significant at the $\alpha = 0.05$ level. However, there are differences between the plots corresponding to all weather surface, dirt surface, and turf surface. Particularly, all weather surface has trends that are different than turf surface and dirt surface. These differences are also seen in previous figures above. The results in Figure 5.8 can be interpreted similarly.

The R Code to create these plots is below:

```
library(reshape2) #loading reshape2 library

# Reshaping the tibble/dataframe from a wide dataset
   to a long dataset
DATA_NEW_melted = melt(DATA_NEW, id.vars = c("Lasix",
    "Surface", "Finish", "Finish Time"),
  variable.name = "Level", value.name="Probability")

## Creating prediction curves with individual
   explanatory variables and their predicted
   probabilities.

library(ggplot2)  #  loading library ggplot2

# Finish Time
ggplot(DATA_NEW_melted, aes(x = `Finish Time`, y =
   Probability, colour = Level)) +
  geom_smooth() +
  ylim(0,0.8)

# Finish
ggplot(DATA_NEW_melted, aes(x = Finish, y =
   Probability, colour = Level)) +
  geom_smooth()+
  ylim(0,0.8)

# Surface
ggplot(DATA_NEW_melted, aes(x = Surface, y =
   Probability, colour = Level)) +
  geom_boxplot()+
  ylim(0,0.8)

# Lasix
```

```
27 ggplot(horse_NEW_melted, aes(x = Lasix, y =
      Probability, colour = Level)) +
28   geom_boxplot()+
29   ylim(0,0.8)
30
31
32 ## Creating prediction curves with multiple
      explanatory variables and their predicted
      probailites.
33
34 # Surface, Lasix and Finish Time
35 ggplot(DATA_NEW_melted, aes(x = `Finish Time`, y =
      Probability, colour = Level)) +
36   geom_smooth() +
37   facet_grid(Surface~Lasix, labeller="label_both")
```

SAS gives these under PROC LOGISTIC PLOTS=all options, so, no special code is needed.

Ordinal Probit Model Example

The example used so far in this chapter applies the ordinal logit model. If the ordinal probit model is desired, the specification of the model would change which would result in the interpretations of the coefficients changing. Otherwise, the other parts of the analyses remains the same. Thus, we show how to fit the ordinal probit model and interpret its coefficients only.

The R Code and the Results for fitting the ordinal probit model is below.

```
1 library(MASS)   ## Loading the package MASS
2
3 model_fit = polr(ordered(ConsensusEIPH) ~ Lasix +
      Surface + Finish + `Finish Time` , data = DATA,
      Hess = TRUE, method = c("probit"))
4 summary(model_fit)
```

Ordered Probit Regression-Log Odds Ratio Estimates

Interpretations

The main difference between using the ordinal logit and ordinal probit models is how the coefficients are interpreted. Instead of interpreting how the logodds ratio corresponding to the response variable changes as the explanatory variables change, we interpret how the z-score corresponding to the response variable changes as the explanatory variables change. The z-score refers to the values on the Standard Normal distribution that measures the number of standard deviations away from the mean the data is. For reference, the

```
Call:
polr(formula = ordered(ConsensusEIPH) ~ Lasix + Surface
+ Finish +
    `Finish Time`, data = DATA, Hess = TRUE, method =
c("probit"))

Coefficients:
                  Value Std. Error t value
LasixY        -0.048507   0.110091 -0.4406
SurfaceDirt    0.826216   0.262052  3.1529
SurfaceTurf    0.813129   0.271358  2.9965
Finish         0.043358   0.012291  3.5277
`Finish Time`  0.005324   0.002178  2.4446

Intercepts:
    Value   Std. Error t value
0|1  1.0361  0.3207      3.2306
1|2  1.9634  0.3231      6.0762
2|3  2.8702  0.3268      8.7814
3|4  3.6738  0.3381     10.8664

Residual Deviance: 2769.477
AIC: 2787.477
```

R Output

FIGURE 5.10
Ordered probit regression estimates

empirical rule for the normal distribution states that 99.7% of the data falls within 3 standard deviations (above or below) the mean.

Interpretation of Coefficients

- Numeric Variables (Continuous and Discrete)
  - As **finish** increases by one unit, the z-score for being more likely to have higher Consensus EIPH increases by 0.043358. This implies that as the finish increases by 1 unit, it is more likely for a horse to have blood in the lungs.
  - As **finish time** increases by 1 unit, the z-score for being more likely to have higher Consensus EIPH increases by 0.005324. This implies that as the finish time increases by 1 unit, it is more likely for a horse to have blood in the lungs.
- Categorical Variables
  - Horses who have **Lasix: yes** has a z score for being more likely to have higher Consensus EIPH that is 0.04857 less than horses that had Lasix: no. This implies that horses who have had the Lasix procedure

were less likely to have blood in their lungs than horses who did not have the Lasix procedure.

- Horses who ran on **surface: dirt** had a z-score for being more likely to have a higher Consensus EIPH that is 0.826216 higher than horses that ran on all weather surface. This implies that horses who run on dirt surface were more likely to have blood in their lungs than horses who ran on all weather surface.
- Horses who ran on **surface: turf** had a z-score for being more likely to have a higher Consensus EIPH that is 0.813129 higher than horses that ran on all weather surface.This implies that horses who run on turf surface were more likely to have blood in their lungs than horses who ran on all weather surface.

Notice that these results and interpretations are similar to the results and interpretations when the ordinal logit model is used. The rest of the analyses is similar to when the ordinal logit model is used.

The SAS code for fitting a generic ordinal probit model is given below:

```
/* ORDINAL PROBIT MODEL*/
PROC PROBIT DATA=dataset;
MODEL Y=X1 X2 X3 X4 ;
run;
```

5.5 Summary

This chapter covered the theory and applications of CDF Models (Ordinal Logit and Probit). Applications include: fitting the model, interpreting coefficients, interpreting confidence intervals and p-values and making predictions based on the model.

As ordinal data is often not talked about explicitly, we decided to provide some of the mathematical detail in terms of what the CDF models and how these models work. We have also provided the dataset and the R and SAS Code, so the readers could analyze the data with us.

6

Latent Variable Models: Structural Equation Models

6.1 Introduction

The Latent Variable Models (LVM), especially Structural Equation Modeling (SEM), is a method that can be used for analyzing ordinal data. Of all the LVM's, SEM is the more closely connected to ordinal data even though it is not strictly for analyzing ordinal data. SEM is more often used when trying to understand and analyze continuous constructs and this has been studied extensively by several statisticians, e.g., Sammel and Ryan (1996) [70], Browne (1984) [15], Bentler (1983) [9], and Andersen (1982) [5].

Like many of the other methods used for ordinal data (e.g., CDF models, Likert scale), there also is an assumption of an underlying latent variable when analyzing ordinal data using LVM. LVM, especially SEM, are used to explain the *interrelationships* among a set of observed variables (continuous, ordinal, or binary) when the observed variables are proxies for multiple latent variables that are thought to exist in the background. For LVM, there are **multiple** latent variables that are used as response **and** explanatory variables. Further, for most LVM, the multiple response variables are not ordinal, but continuous. For example, for the latent trait intelligence, several observed variables like IQ scales (which are continuous) can be used. One important caveat about SEM is that while multiple latent variables are allowed in the model for purpose of estimability, the number of latent variables is typically assumed to be much smaller than the number of observed variables.

Latent variable models with observed ordinal variables are particularly useful for analyzing survey data and this is the place where it intersects with ordinal data. Often, as we saw in Likert's seminal paper ([51]), ordinal variables are used to understand the underlying unobserved construct through a series of attitudinal questions about their degree of agreement or liking. As we have seen before, alternatives like "strongly disagree", "disagree", "strongly agree" or "very dissatisfied", "dissatisfied", "satisfied", and "very satisfied" are often used to understand the latent construct. In a Likert type analysis, we use

DOI: 10.1201/9781003020615-6

several such statements to understand a latent construct. In CDF methods we use one statement and use the observed *non-stochastic* explanatory variables to understand the latent variable. SEM on the other hand, assumes that there are multiple latent variables both in the response and explanatory variables. Hence, some important distinctions between a classic CDF and SEM models are:

- SEM have potentially multiple latent responses
- The explanatory variables are allowed to be measured with error (stochastic)
- There are multiple equations connecting explanatory and response to other responses
- One can use linear and generalized regression for analysis incorporating Confirmatory Factor Analysis and path analysis as tools

There are two main approaches for analyzing ordinal data based on latent variables. The most popular one is the Underlying Variable Approach (UVA), (see Muth'en (1984) [62], J'oreskog (1990) [39] which assumes that the observed variables are generated by underlying normally distributed continuous variables). This approach is used in structural equation modeling (SEM) and the relevant methodological developments are available in commercial software such as LISREL (J'oreskog and S'orbom (1988) [41]) and Mplus (Muth'en and Muth'en (1998–2012) [64]). The other approach is the Item Response Theory (IRT), where unit of analysis is the entire response pattern of a subject. An overview of those type of models can be found in Bartholomew and Knott (1999) [48] and van der Linden and Hambleton (1997) [83]. Moustaki and Knott (2000) [61] and Moustaki (2000) [60] discuss a Generalized Linear Latent Variable Model framework (GLLVM) for fitting models with different types of observed variables.

In this book, we will focus on the SEM models as opposed to IRT and focus on looking at techniques that have available R packages to illustrate our examples. We acknowledge existence of studies on IRT like J'oreskog and Moustaki (2001) [40], Cagnone et al. (2005) [16] which show that the IRT approach is preferable in terms of accuracy of estimates and model fit. However, as this book is focused on the analysis and prediction using a single ordinal response and not on testing theory we do not talk about these methods in this book. For interested readers we refer to recently proposed techniques by Schilling and Bock (2005) [71]. In many occasions, it is assumed that an ordinal variable is a result of the discretization of a continuous latent variable (Bock (1975) [12]).

6.2 Structural Equation Modeling

SEM is a collection of quantitative techniques that is used to test and estimate complex relationships between variables. It's often used in social sciences, psychology, economics, and other fields to analyze the relationships among latent (unobserved) constructs and observed variables. SEM comes into play when dealing with variables that cannot be directly measured with precision. Instead, sets of items are used that represent these underlying concepts. For example, measuring "happiness". There is no "happiness meter" that can measure this for us. Hence, we say it is not directly observable. However, we can use questions about well-being, satisfaction, and mood to get an idea of someone's happiness level. In SEM, underlying concepts, like happiness, are called latent variables or factors. They are like hidden traits that influence the observed items.

To explain these methods and their nuances, we first set up some notation and variables while using the language of both classical statistics and educational psychology. We do want to emphasize that we use the notations that are more conventional in Statistics.

- x denotes a single explanatory variable. This, in classical statistics, is assumed to be non-stochastic and measured without any error. In Economics, this is often called the exogenous variable. In Education literature, this is referred to as the observed variable. As defined here, x can be numerical, ordinal or categorical.
- $x_1, ..., x_n$ be a random sample of observations for the explanatory variable x.
- $\mathbf{x}$ denotes multiple explanatory variables, the **bold** denoting a matrix, so that there are multiple explanatory variables.
- Y denote the single response variable, which is stochastic i.e., measured with error, so is considered a random variable. As defined before, Y can be numerical, ordinal, or categorical. In Economics, this is referred to as endogenous variable.
- Similarly, $\mathbf{Y}$ denotes multiple responses or a vector of responses.
- η is a latent variable. In models like CDF models, η is necessarily a **response** variable that is unobserved but can be analyzed via the observed x. In Likert models, it can be constructed using a summation of ordinal observed variables. In some SEM models we could have both explanatory (exogenous) and response (endogenous), η.
- α and β denote the parameter vectors for intercept and slope, respectively for modeling exogenous to endogenous variables, or predicting Y using x.

- γ are the parameters relating endogenous to endogenous variables
- λ are the parameters relating observed variables to latent variables (factor loadings).
- $\boldsymbol{\beta}$ represents the matrix of slope parameters between multiple responses $\mathbf{Y}$ and $\mathbf{Y}$.
- $\boldsymbol{\Gamma}$ represents the matrix of parameters between the responses.
- $\boldsymbol{\Lambda}$ represents the matrix of factor loadings.
- J denotes the column vector of ones, $(1, 1...1)^t$ used to incorporate the intercept into the model.
- Let ϵ represent the random error vector
- $\boldsymbol{\epsilon}$ represent the matrix of random errors

To note, all matrices are **bold** and all vectors are kept unbold.

6.3 The Models

SEM is a collection of methods relating x, Y, and η. Out of the following subsections only 6.3.5 and 6.3.6 deal with potentially ordinal data and are of interest to us obliquely. We include the other sections for statistical clarity on these topics.

6.3.1 Simple Linear Regression: Relating One X to One Y

Here, we model one numerical response variable as being related to one numerical explanatory variable.

$$\textbf{Model: } Y_i = \alpha + \beta x_i + \epsilon_i$$

An example is trying to explain scores on a Math final exam (Y_1) using midterm grades (x_1) as the predictor.

The R and SAS Code is given below. Since there is not a motivating dataset, the code provided is general. For consistency we are using the **lavaan** package in R for all the models, though one could do most of these models using much simpler functions like **lm()**. The **lavaan** package is used to fit latent variable models, including confirmatory factor analysis, SEM and latent growth curve models. Similarly we are using **PROC CALIS** in SAS for all models even though we could use other PROC as well.

```
library(lavaan)  # loading lavaan package

# Specify Simple Linear Regression
m1 =
    y1 ~ 1 + x1
,
fit1 <- sem(m1, data=dat)
summary(fit1)
```

As we mentioned earlier that in R where the package LAVAAN is used, in SAS the PROC used in SEM is PROC CALIS. The symbol "===>" indicates relationship in PROC CALIS. The relationship between Y and X is given by "PATH" instead of "MODEL" in PROC REG.

The SAS code is given below:

```
/* SIMPLE LINEAR REGRESSION*/
PROC CALIS DATA=dataset;
   PATH   X1   ===>  Y;
run;
```

6.3.2 Multiple Linear Regression: Relating Multiple X's to One Y

Here, we model one numerical response variable as being related to multiple explanatory variables (continuous, discrete or categorical).

Written often in the matrix notation and called **General Linear Model**.

$$\textbf{Model: } Y = J\alpha + \mathbf{X}\beta + \epsilon$$

with β representing a vector of parameters.

To note: this notation separates the intercepts from the slopes, unlike the usual notation of a GLM.

An example is trying to explain scores on Math final (Y_1) exam using midterm grades (x_1), homework grades, and attendance (x_2) as the predictors.

The R Code is below:

```
library(lavaan)     # loading lavaan package

# Specify Multiple Linear Regression
m2 <-
```

```
    y1 ~ 1 + x1 + x2
'
fit2 <- sem(m2, data=dat)
summary(fit2)
```

The SAS code is given below:

```
/* MULTIPLE REGRESSION*/
PROC CALIS DATA=dataset;
   PATH   x1 x2   ===>  y;
run;
```

In both these models we have one endogenous Y (response) with one or more exogenous (predictors) that are measured without error. The entire randomness in the model is from the response and not the predictors.

- Sections 6.3.1 and 6.3.2 discuss regression models that are only loosely related to SEM

6.3.3 Multivariate Regression: Relating Multiple X's to Multiple Y's

In this case we model multiple responses (or endogenous variables) with observed non-stochastic or exogenous variables.

$$\textbf{Model: } \mathbf{Y} = J\boldsymbol{\alpha} + \mathbf{x}\boldsymbol{\beta} + \boldsymbol{\epsilon}$$

Here $\mathbf{Y}$ represents the response matrix, for multiple predictors. The matrix $\mathbf{x}$ is our design matrix of predictors. The matrices $\boldsymbol{\alpha}$ and $\boldsymbol{\beta}$ are used for the parameters. Finally, $\boldsymbol{\epsilon}$ is the matrix of random errors and J is the matrix of 1's.

For modeling two endogenous variables with multiple exogenous x, we have:

$$\begin{bmatrix} Y_1 & Y_2 \end{bmatrix} = J \begin{bmatrix} \alpha_1 & \alpha_2 \end{bmatrix} + \mathbf{x} \begin{bmatrix} \beta_1 & \beta_2 \end{bmatrix} + \begin{bmatrix} \epsilon_1 & \epsilon_2 \end{bmatrix} \tag{6.1}$$

An example is trying to explain scores on Math final (Y_1) and Physics final (Y_2) exams using midterm grades for Math (x_1) and Physics (x_2), homework grades for Math (x_3) and homework grades for Physics (x_4) and attendance (x_5) as the predictors.

The R Code is below:

```
library(lavaan)    # loading lavaan package

# Specify Simple Linear Regression
```

```
m3 <-
    y1 ~ 1 + x1 + x2 + x3 + x4 + x5
    y2 ~ 1 + x1 + x2 + x3 + x4 + x5
'
fit3 <- sem(m3, data=dat)
summary(fit3)
```

The SAS code is given below:

```
/* MULTIVARIATE REGRESSION*/
PROC CALIS DATA=dataset;
   PATH    x1 x2 x3 X4 X5   ===>  y1;
   PATH    x1 x2 x3 X4 X5   ===>  y2;
run;
```

6.3.4 Path Analysis: Endogenous to Endogenous, Relating Response Y's to Each Other

The difference between these models and the ones we talked about in the previous sections is the fact that the predictor variables here (we do not call them explanatory and denote them as x purposely) are themselves measured with error. In econometrics terms we would call it, endogenous response and potentially exogenous predictors.

In matrix form, the model would be written as:

$$\textbf{Model: } \mathbf{Y} = \mathbf{J}\boldsymbol{\alpha} + \mathbf{x}\boldsymbol{\beta} + \mathbf{Y}\boldsymbol{\gamma} + \boldsymbol{\epsilon}$$

Here $\mathbf{Y}$ represents a matrix of response or multiple responses and can be both predictors and response, for a myriad of $\mathbf{x}$ and a matrix of parameters, $\boldsymbol{\alpha}$ and $\boldsymbol{\beta}$ and $\boldsymbol{\epsilon}$ matrix of errors. Here, the added term $\boldsymbol{\gamma}$ are the parameters from the endogenous factors.

Theoretically, this could be a more difficult problem, however, we make assumptions that allows the estimation of the parameters. The assumptions are:

- $E(\epsilon) = 0$
- $\boldsymbol{\epsilon}$ and $\mathbf{x}$ are uncorrelated
- Matrix $(I - \gamma)$ is invertible.

An example is trying to explain scores on Math final (Y_1) and Physics final (Y_2) exams using midterm grades for Math (x_1) and Physics (x_2), homework grades for Math (x_3), homework score for Physics (x_4), and attendance (x_5) as the predictors, however we believe that the scores on the Physics exam is explained by the Math final exam score. This last belief adds stochastic term(s) to the predictors.

The R Code is below:

```
library(lavaan)      # loading lavaan package

# Specify Structural Model
m4 <-
     y1 ~ 1 + x1 + x2 + x3 + x4 + x5
     y2 ~ 1 +  x2  + x4 + x5 +y1
,
fit4 <- sem(m4, data=dat)
summary(fit4)
```

The SAS code is given below:

```
/* MEASUREMENT ERROR MODEL*/
PROC CALIS DATA=dataset;
    PATH   x1 x2 x3 x4 x5  ===>  y1;
    PATH   x2 x4 x5 y1  ===>  y2;
run;
```

- Till this point all our response variables were numerical and directly measured
- In the next two subsections, it will be clear why SEM is included in this book, as it is really the only LVM that is related to ordinal data, even if obliquely.

6.3.5 Confirmatory Factor Analysis: Relating Latent Response(s) with Observed x

Instead of using direct measures, let us consider a scenario where we are interested in "motivation" which is not directly measurable. We believe that this latent variable "motivation" could be an important predictor in the overall performance of the student. While not directly measurable, there is a strong belief that we can ask a series of questions to understand this latent variable, which can then help us build models with this latent variable as a predictor or as a response (depending upon the scenario). These could be done using a survey asking questions on "motivation". We are assuming that there is a such a survey that is possible and has all the features of a good survey (internal consistency, etc., which is not something we will discuss). As these are survey questionnaires the response to the surveys are often ordinal. And **herein lies**

our interest in this topic. The process of "estimating" the latent construct is done through Confirmatory Factor Analysis.

To understand CFA, we will first briefly talking about Factor Analysis and what it involves. If we consider a simple linear regression model, our model is:

$Y = J\alpha + \mathbf{X}\beta + \epsilon.$

And in this scenario our Y and our x are fully observed. In Factor Analysis, the idea is that our Y is not observed but can be constructed using the X. This process of construction is done by de-constructing the observed Variance-Covariance Matrix. The general tenet for FA to work is the belief that the observed x is composed of observations that are correlated with each other, and this correlation is due to the fact they all explain the underlying unobserved construct. Let us look at the simplest case where we have three observed x_1, x_2, x_3 all related to an unobserved η.

Hence, the idea is instead of a simple linear regression written as:

Model: $Y_i = \alpha + \beta_1 x_1 + \beta_2 x_2 + \beta_3 x_3 + \epsilon$

However, we do not know Y as we have a latent construct η and technically

Model: $\eta = \alpha + \beta_1 x_1 + \beta_2 x_2 + \beta_3 x_3 + \epsilon$

We will write this in terms of the observed x and the latent

Model: $x_1 = \tau_1 + \lambda_1 \eta + \delta_1$

Model: $x_2 = \tau_2 + \lambda_2 \eta + \delta_2$

Model: $x_3 = \tau_3 + \lambda_3 \eta + \delta_3$

The simplistic way to explain this is considering $\Sigma(\Theta)$ to the variance Covariance matrix of the $x's$ where,

$\Sigma(\Theta) = \Lambda \Psi \Lambda^t + \Theta_\delta$

Further,

- Λ is factor loading matrix
- Ψ variance-covariance matrix of the latent factors
- θ_ϵ variance-covariance matrix of the residuals

The general idea is trying to estimate the λ parameters that give us the loadings of the various x on the latent η.

In CFA, we believe we *know* the direct measurements that contribute to the latent variable in question. However, since it is easy for the model to have too many parameters, as the η are unknown, often the off diagonal elements of Θ_ϵ is assumed to be 0.

In summary, CFA is used to assess how well observed variables (indicators) reflect latent constructs. It involves specifying the relationships between latent constructs and their indicators. For example, suppose we are studying intelligence and want to measure it using three observed variables: verbal ability, mathematical ability, and logical reasoning. These observed variables can be indicators of a latent construct called "intelligence".

$$\begin{aligned} intelligence &= \lambda_1 * verbal + \lambda_2 * math + \lambda_3 * logical \\ verbal &= \lambda_4 * intelligence + \delta_1 \\ math &= \lambda_5 * intelligence + \delta_2 \\ logical &= \lambda_6 * intelligence + \delta_3 \end{aligned}$$

The R Code is below:

```
library(lavaan)      # loading lavaan package

# Specify CFA model
cfa_model <- '
    intelligence =~ verbal + math + logical
    verbal =~ intelligence
    math =~ intelligence
    logical =~ intelligence
'

# Fit the model
cfa_fit <- sem(cfa_model, data=your_data)
summary(cfa_fit)
```

The SAS code is given below:

```
/* CFA */
PROC CALIS DATA=dataset;
   FACTOR
      XFACT $===>$ x1-x3,
      YFACT   $===>$ y1-y2;
   PVAR
      XFACT = 1.,
      YFACT = 1.;
RUN;
```

6.3.6 Structural Equations: Latent to Latent Relationship

The sections 6.3.1 to 6.3.5 talk about individual elements of structural regression. Structural Regression Models are the combination of all these models. The real contribution of SEM comes from exploring the relationships between these latent factors. Here our data has observed response, latent response, observed explanatory, and latent explanatory variables.

$$x = J\tau_x + \mathbf{X}\Gamma_x + \delta$$

$$\eta = J\alpha + \mathbf{X}\beta + \eta\Lambda + \zeta$$
$$Y = J\tau_y + \eta\Gamma_y + \epsilon$$

Let us consider an example where We want to study the relationship between "Parenting Style", "Child Behavior", and "Child Academic Performance". We will use structural equation modeling to examine how these variables interact. Often, these Structural Equation Models are represented using path diagrams to give a graphical representation of the models with the errors, observed, and latent variables.

- We hypothesize that the parenting style adopted by parents affects both the behavior of their children and their academic performance.
- The path diagram illustrates the relationships among variables in the context of the example involving "Parenting Style", "Child Behavior", and "Child Academic Performance".

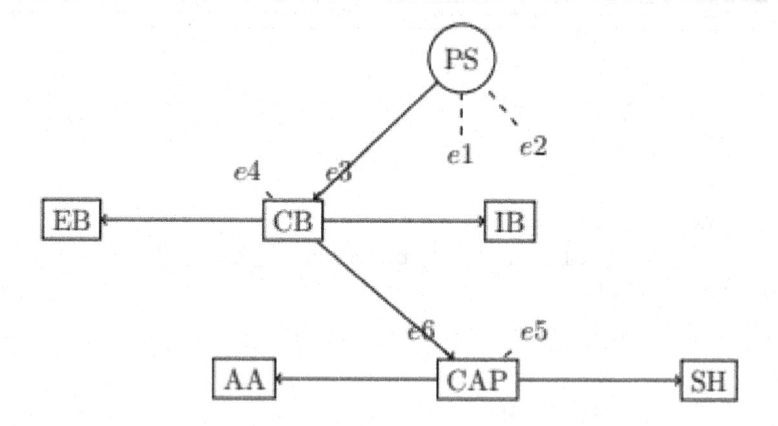

- Parenting Style (PS): This latent variable represents the parenting style adopted by parents. It directly influences both Child Behavior (CB) and Child Academic Performance (CAP).
- Child Behavior (CB): This latent variable represents the behavior of children. It is influenced by Parenting Style (PS) and has two observed indicators: Externalizing Behavior (EB) and Internalizing Behavior (IB).

- Child Academic Performance (CAP): This latent variable represents the academic performance of children. It is influenced by Child Behavior (CB) and has two observed indicators: Academic Achievement (AA) and Study Habits (SH).
- Arrows: The arrows between variables represent causal pathways. For example, the arrow from Parenting Style (PS) to Child Behavior (CB) indicates that parenting style influences child behavior.
- Error Terms: The dashed lines represent error terms associated with each variable. These error terms account for unexplained variance in the variables, which cannot be attributed to the modeled relationships.

In summary, the path diagram visually depicts how parenting style influences child behavior, how child behavior impacts academic performance, and how these relationships are captured in a structural equation model. Once we have the path diagram, let us understand the measurement model equations.

- Parenting Style (PS) measured by indicators Strictness (S) and Warmth (W):
  $S = \lambda_1 * PS + \delta_1, W = \lambda_2 * PS + \delta_2$
- Child Behavior (CB) measured by indicators Externalizing Behavior (EB) and Internalizing Behavior (IB):
  $EB = \lambda_3 * CB + \delta_3, IB = \lambda_4 * CB + \delta_4$
- Child Academic Performance (CAP) measured by Academic Achievement (AA) and Study Habits (SH):
  $AA = \lambda_5 * CAP + \delta_5, SH = \lambda_6 * CAP + \delta_6$

Now, we will formulate the structural model equations.

- Relationship between Parenting Style and Child Behavior:
  $CB = \beta_1 * PS + \zeta_1$
- Relationship between Child Behavior and Child Academic Performance:
  $CAP = \beta_2 * CB + \zeta_2$
- A positive β_1 value would suggest that a more strict parenting style is associated with certain behaviors in children. A positive β_2 value would indicate that specific child behaviors are linked to better academic performance. The error terms (u and e values) capture unexplained variance in the variables.

For our example, we would model it as:

$$academic\ motivation = a * parental\ involvement + \zeta_1$$
$$academic\ achievement = b * academic\ motivation + \zeta_2$$

The R Code is below:

```
library(lavaan)        # loading lavaan package

# Specify path analysis model
path_model <- '
    academic_motivation ~ a * parental_involvement
    academic_achievement ~ b * academic_motivation
'

# Fit the model
path_fit <- sem(path_model, data=your_data)
summary(path_fit)
```

We provide a generic SAS code after the next example.

The Path Analysis examines the relationship between academic motivation, parental involvement, and academic achievement. The coefficient a represents the direct effect of parental involvement on academic motivation. For every one-unit increase in parental involvement, we expect a change of a in academic motivation. The coefficient b represents the direct effect of academic motivation on academic achievement. For every one-unit increase in academic motivation, we expect a change of b in academic achievement. The error terms (ζ_1, ζ_2) capture unobserved factors affecting the variables. (Note: The interpretation is simplified and does not account for possible endogeneity or omitted variable bias.)

Full Structural Equation Modeling (Full SEM): Full SEM combines measurement models (for latent constructs) and structural models (for relationships among variables), allowing for a comprehensive analysis of complex relationships.

$$\begin{aligned} latent\ parental &= \lambda_1 * parentalinvolvement \\ latent\ motivation &= \lambda_2 * academic\ motivation \\ academic\ achievement &= c * latent\ parental + d * latent \\ &\quad motivation + e * SES + \epsilon_1 \end{aligned}$$

The R Code is below:

```
library(lavaan)      # loading lavaan package

# Specify full SEM model
full_sem_model <- '
    # Measurement models
    latent_parental =~ parental_involvement
    latent_motivation =~ academic_motivation

```

```
    # Structural model
    academic_achievement ~ c * latent_parental +
    d * latent_motivation + e * SES
'

# Fit the model
full_sem_fit <- sem(full_sem_model, data=your_data)
summary(full_sem_fit)
```

The SAS code is given below:

```
/*PATH EQUATIONS*/
PROC CALIS DATA=dataset;
   LINEQS
      x3 = alpha1 * Intercept + beta1  * x1  + gamma1
    * x2 + E1,
      y = theta1 * Intercept + theta2 * x3  + theta3
   * x1 + theta4 * x2 + E2;
   VARIANCE
      E1-E2 = eps1-eps2;
   COV
      E1-E2 = eps3;
   BOUNDS
      eps1-eps2 > 0. ;
RUN;
```

The Full SEM combines measurement models for latent constructs (latent parental, latent motivation) and a structural model for relationships. The coefficients λ_1 and λ_2 represent the factor loadings between latent constructs and their indicators. The coefficient c indicates the effect of latent parental on academic achievement, while d indicates the effect of latent motivation on academic achievement. The coefficient e represents the effect of SES on academic achievement. The error terms (ϵ) capture unexplained variance. (Note: The interpretation is simplified and does not account for potential endogeneity or omitted variable bias.)

6.4 Structural Equation Modeling for Ordinal Data

In most of the literature on SEM for ordinal data, the key concepts of latent variables, measurement models, and structural relationships remain the same as in traditional SEM. However, the focus shifts to appropriately modeling the ordered categorical indicators. For an ordinal indicator Y_i with L ordered categories, the probability of observing category j is modeled using the cumulative probability function:

$$P(Y_i = j) = P(\tau_{j-1} < F < \tau_j)$$

Where as in chapter 5, τ represents thresholds, and F is the underlying latent variable. This captures the probability of the latent variable falling within a certain range defined by the thresholds. The methods of relating an observed variable to an ordinal response has been addressed in chapter 5.

Several estimation methods can handle SEM for ordinal data:

- WLSMV (weighted least squares) is a widely used method for SEM with ordinal data. It takes into consideration the categorical nature of the variables by estimating the polychoric correlations or tetrachoric correlations. WLSMV adjusts the model estimation based on the means, variances, and covariances of the ordinal indicators. It provides robust parameter estimates (robust to non-normality), factor loadings, thresholds, and relationships. It's available in software packages like "Mplus" and the "lavaan" package in R.

- DWLS (Diagonally Weighted Least Squares): DWLS is similar to WLSMV and is suitable for handling ordinal indicators. It focuses on the diagonal elements of the covariance matrix, making it computationally efficient for large datasets. It is also robust to non-normality.

- MLR (Robust Maximum Likelihood): Robust MLR is a method that can be used for both continuous and ordinal data. It assumes multivariate normality of the indicators and incorporates robust standard errors to account for non-normality. While not specifically designed for ordinal data, it can still be applied effectively with appropriate considerations.

R code to run an SEM for ordinal data using WLSMV estimation method is given below:

```
1  library(lavaan)    # loading lavaan package
2
3  # Specify measurement model
4  measurement_model <- '
5     ability =~ ability_item1 + ability_item2 +
    ability_item3
6     performance =~ performance_item1 + performance_
    item2 + performance_item3
7  '
8
9  # Specify structural model
10 structural_model <- '
```

```
11      performance ~ beta * ability
12  '
13
14  # Combine models
15  full_model <- paste(measurement_model, structural_
        model)
16
17  # Fit the model using WLSMV
18  sem_fit <- sem(full_model, data=your_data, estimator
        = "WLSMV")
19  summary(sem_fit)
```

6.5 Summary

In this chapter we gave a brief statistical presentation of SEM models and discuss why these are not relevant for content of the book. Ordinal data in SEM are generally observed and converted into a continuous latent variable via CFA. If the observed response itself is ordinal, we can use methods in Chapter 5 for analysis. While, we gave an idea of how to do SEM using R, we did not go into details of the methods as we did in chapter 5, as these are not relevant methods for analyzing ordinal responses. In general, SEM methods use the ordinal observations to create their latent variables as inputs in their models. There is not an emphasis of analyzing the ordinal response as a function of predictors and understanding the single ordinal response.

Part IV

Further Work with Ordinal Data

7

Diagnostics of the Ordinal Regression Models

7.1 Introduction

Diagnostics as the word denotes, comprises of methods and techniques to diagnose "issues" with the model. Questions diagnostics can answer include: (1) How well does the model represent the data? and (2) Are there some assumptions that are being violated? As we are focusing for the most part on ordinal logistic regression, we will look at diagnostics from the point of view of this model. If the probit model is used, similar diagnostics can be done as well.

These questions are easier to answer for numerical data, where we look to see if the assumptions made for analysis (under the General Linear Model, GLM) like uncorrelatedness, equality of variance, and distributional assumptions (Normal distribution) are satisfied. Diagnostics is not so clear cut when the data are not numerical or the models used are not GLM or related models.

One major issue when performing diagnostics is the lack of an additive error in the model statement. This makes it difficult to measure the errors. For example, consider the model statement for GLM:

$$Y = X\beta + \epsilon$$

with the usual assumptions that the error. ϵ, follows a Normal Distribution with parameters (mean $= 0$, standard deviation $= I\sigma$)

When using a GLM, the residuals for the model are used to estimate ϵ and diagnostics are performed on the residuals to check if the assumptions are indeed satisfied.

Now, consider the models for the ordinal regression models. From chapter 5, CDF model is of the form:

$$\Gamma^{-1}[P(Y \leq j|X)] = \alpha_j - \beta^T X, \qquad j = 1, \ldots, c-1$$

DOI: 10.1201/9781003020615-7

where Γ is the CDF of the underlying latent variable. If Γ is the Normal CDF we have the probit models, if we consider Γ to be the logistic CDF we have the logistic model. In all generalized linear models, the assumption of additive error is no longer an option as the relationship between the response and the explanatory variable is not additive. As shown in chapter 5, a function of the CDF (or cumulative probability) related to x, so the random error of the system is not directly observable by the residuals.

Understanding Residuals

We will show briefly how residuals in the General Linear Models (GLM) and the CDF models are different entities. In GLMs, we assume that our response Y is a continuous random variable, $Y = X\beta + \epsilon$, with the assumption that ϵ follows a Multivariate Normal Distribution, with mean vector 0 and variance covariance matrix $I\sigma$, where I is the identity matrix. Thus the implicit assumptions of the model are normality, equal variance, and uncorrelatedness among the response.

We estimate, β by using MLE or Ordinary Least Squares and our estimate is: $\hat{\beta} = (X'X)^{-1}X'Y$. We predict our response Y by $\hat{Y}$ which is given by $X\hat{\beta}$. Thus, the residuals, $\mathbf{e}$ is given by $\mathbf{e} = Y - \hat{Y}$. The residual is then used as a proxy for the random error ϵ.

In contrast the CDF models, or models in the framework of GLiM, relate to a response Y that is **NOT** continuous. Binary Logit or Probit deal with Y as a binary random variable and is modeled using Binomial distribution. For ordinal logit we use the multinomial or Binomial distribution. The point to keep in mind is in GLiM, the relationship is not between Y and X directly. The GLiM is written in terms of the Expected Value of Y as a function of $X\beta$:

$$E(Y) = F(X\beta).$$

In the binary or ordinal set-up, we assume Y follows a Binomial distribution. Hence, $E(Y)$ is the probability of success or probability of being in a certain category. The functional form $F(.)$ is the CDF function. Hence, there are no direct residuals as in the GLM model. $\hat{Y}$ is a stage and so $e = Y - \hat{Y}$ is not a continuous random variable and not defined as in GLM. For binary response, the residual can only take values of -1, 0, or 1. Similarly in ordinal regression with 5 categories, it can take values from -4 to 4.

7.2 Assumptions for Ordinal Logistic Model

To do diagnostics on Ordinal CDF models, we need to first understand what our assumptions are for these models. Our assumptions are:

- The correct model is specified.
- The proportional odds assumptions are satisfied.
- The response variables are uncorrelated.
- The explanatory variables are not correlated among themselves.

To understand whether the correct model is specified, we look at Goodness of Fit tests, where we look at the observed values and compare it to the expected values under the assumption of the model. In the next section we will look at a few different Goodness of Fit tests that are relevant for ordinal data.

The proportional odds assumption is a fairly central assumption for the ordinal logistic regression. It essentially says that the slope of the x variables do not change across the different categories of the ordinal variable. Hence, essentially it allows the estimation of the parameters without running out of degrees of freedom.

Mathematically, the proportional odds model is:

$$logitP(Y \leq j) = \alpha_j - x_i\beta$$

as opposed to

$$logitP(Y \leq j) = \alpha_j - x_i\beta_j$$

or the category j changing the slope parameter. This is also called the parallel lines assumption as shown in chapter 5.

If the response variables are themselves correlated, then we violate the assumption of uncorrelatedness among the Y. The most common type of correlation is autocorrelation. To test for autocorrelation we do need the order of data collection or have a time/space information for the response. For ordinal response, where the observations are quasi-numerical it is a hard assumption to check for.

If the explanatory variables are themselves correlated, we say we have an issue of multi-collinearity among the predictors. Unlike, the other assumptions, it is related to the explanatory x, not the ordinal Y. Hence, here methods used are similar to methods used in GLM.

To understand whether the model fits the data well, we revert back to the Goodness of Fit statistics from 1900. In conjunction with this we will use Association tests, to see how the predicted and observed are associated. We will also explore some Model Fit criterion using Likelihood.

7.3 Goodness of Fit Tests

For these tests, we are using the null hypothesis that the model fits the data well. Hence, rejecting the null hypothesis implies that the model does not fit the data well for some unspecified reason. These types of diagnostics are called Goodness of Fit tests. In this section we discuss a few commonly used Goodness of Fit Tests.

7.3.1 Hosmer and Lemeshow Test

The ordinal version of the Hosmer-Lemeshow (HL) test is derived from the multinomial and binary versions of the test. Hosmer and Lemeshow (1989) [35] proposed a Pearson statistic comparing the observed to expected counts within each group in a $g \times c$ contingency table. The ordinal HL test statistic is given by:

$$C_g = \sum_{k=1}^{g} \sum_{j=1}^{c} \frac{(O_{kj} - \hat{E}_{kj})^2}{\hat{E}_{kj}}$$

where O_{kj} is the sum of the observed frequencies in each g group for each response category and $\hat{E}_{kj}$ is the sum of the estimated frequencies in each group for each response category. These are derived using the following equations:

$$O_{kj} = \sum_{i \in \Omega_k} \tilde{y}_{ij},$$

$$\hat{E}_{kj} = \sum_{i \in \Omega_k} \hat{\pi}_{ij}$$

where $\tilde{y}_{ij}$ is a binary indicator variable with $\tilde{y}_{ij} = 1$ when $y_i = j$ and $\tilde{y}_{ij} = 0$ otherwise, and Ω_k denotes the set of indices of the n/g observations in group k.

This statistic follows a Chi-squared distributed with $(g-2)(c-1)+(c-2)$ degrees of freedom under a correctly fit model(Fagerland and Hosmer(2007) [25], Ugba (2020) [82]). If the overall deviation of observed frequencies from the expected frequencies is large, then the statistic will be large and the p-value will be small, resulting in rejection of the null hypothesis that the data arise from the specified model. In other words, a large p-value indicates a good model fit and a small p-value indicates a poor model fit.

The R Code and result is below. This test illustrated using the ordinal logit model (model_fit) from chapter 5.

```
1 library(gofcat) # # Loading the package gofcat
2
3 # Hosmer-Lemeshow Test
4 hosmerlem(model_fit,tables = TRUE)
```

```
Hosmer-Lemeshow Test:

                      Chi-sq  df  pr(>chi)
ordinal(Hosmerlem)    42.227  35     0.187

H0: No lack of fit dictated

rho: 98%
```

Output with Hosmer and Lemeshow test results

FIGURE 7.1
Hosmer-Lemeshow test results

The results show a chi-squared statistic of 42.227 on 35 degrees of freedom. This corresponds to a p-value of 0.187. In this example, our α is 0.05. Since the p-value is greater than the desired $\alpha = 0.05$, the null hypothesis is not rejected. This implies that there is not enough evidence to show that there is not a lack of fit. In other words, we conclude that the model adequately fits the data.

7.3.2 The Pulkstenis-Robinson (P-R) Test

The Pulkstenis-Robinson (P-R) tests are capable of accommodating models with continuous as well as categorical predictors. The observations are grouped according to observed covariate patterns using the categorical covariates only. The covariate patterns are split into two at the median score within each pattern (Ugba (2020) [82]). The P-R test statistic are the Pearson chi squared and deviance statistics on the contingency table formed from tabulating covariate patterns with response categories

$$PR(\chi^2) = \sum_{l=1}^{2}\sum_{k=1}^{K}\sum_{j=1}^{c}\frac{(O_{lkj} - \hat{E}_{lkj})^2}{\hat{E}_{lkj}}$$

and

$$PR(D^2) = 2\sum_{l=1}^{2}\sum_{k=1}^{K}\sum_{j=1}^{c} O_{lkj} log\frac{O_{lkj}}{\hat{E}_{lkj}}$$

where l indexes the two subgroups based on the ordinal score, K is the number of observed covariate patterns because of the categorical covariates, and c is the number of response categories. Both $PR(\chi^2)$ and $PR(D^2)$ are distributed by the chi-squared distribution with $(2K - 1)(c - 1) - p_{cat} - 1$ degrees of freedom, where p_{cat} denotes the number of dichotomous variables needed to model all the categorical covariates (Fagerland and Hosmer (2017) [25]). Like the HL test, a large p-value indicates an adequate model fit.

The R Code and result is given below. Again, this test is illustrated using the ordinal logit model from chapter 5.

```
library(gofcat) # # Loading the package gofcat

# Pulkstenis-Robinson Test
pulkroben(model_fit, tables = TRUE)
```

```
Pulkstenis-Robinson Chi-squared Test:

                     chi-sq   df   pr(>chi)
ordinal(pulkroben)   62.657   41     0.0163 *
---
Signif. codes:
0 '***' 0.001 '**' 0.01 '*' 0.05 '.' 0.1 ' ' 1

H0: No lack of fit dictated

rho: 68.33%


Observed Freqs:


Expected Freqs:


 p <= median OS; q > median OS

Warning: Less than 80% of the estimated frequencies are
greater than 1, test results may be inaccurate.
```

Output with Pulkstenis-Robinson test results

FIGURE 7.2
Pulkstenis-Robinson test results

The results show that the chi-squared statistic is 62.657 on 41 degrees of freedom. This corresponds to a p-value of 0.0163, which is smaller than the desired α of 0.05. This means that we can reject the null hypothesis and conclude that there is enough evidence to conclude lack of fit. This result is not consistent with the result of the Hosmer and Lemeshow test.

7.3.3 The Lipsitz Test

The Lipsitz test involves binning the observed data into equally sized g groups based on an ordinal response score. This score is computed by summing the predicted probabilities of each subject for each outcome level multiplied by

```
Lipsitz Test:

                      LR    df    pr(>chi)
ordinal(lipsitz)   14.23     9      0.1144

H0: No lack of fit dictated
```

FIGURE 7.3
Lipsitz test results

equally spaced integer weights (Ugba (2020) [82]). Given this partitioning of the data, the dummy variables below are created.

$$I_{ik} = \begin{cases} 1 & \text{if observation } i \text{ is in group } k \\ 0 & \text{otherwise} \end{cases} \tag{7.1}$$

for $i = 1....,n$ and $k = 1,...,g-1$. The ordinal model is then refit with these dummy variables:

$$g_j(\mathbf{x}) = \alpha_j \pm \beta'(\mathbf{x}) + \gamma_1 I_1 + ... + \gamma_{g-1} I_{g-1} \quad j = 1,...,c-1$$

If the original fit model has good fit, $\gamma_1,...,\gamma_{g-1} = 0$. The likelihood ratio statistic $-2(L_1 - L_0)$ can be used to compare the loglikelihood of the model with (L_0) and without (L_1) the dummy variables. The observed value of the test statistic can be compared with the chi-squared distribution with $g-1$ degrees of freedom (Fagerland and Hosmer (2017) [25]). The null hypothesis for the Lipsitz test is the same as the Hosmer-Lemeshow and P-R test and hence the result is interpreted similarly.

The R Code and result is below. Here, the Lipsitz test is illustrated using the ordinal logit model from chapter 5.

```
library(gofcat) # # Loading the package gofcat
# Lipsitz Test
lipsitz(model_fit)
```

The Lipsitz test results in a likelihood-ratio statistic of 14.23 on 9 degrees of freedom and p-value of 0.1144. Because the p-value is greater than the chosen $\alpha = 0.05$. Like the HL test, the Lipsitz test result imply that the model adequately fits the data because we do not have enough evidence to reject the null hypothesis of good model fit.

We notice that the three goodness of fit tests did not all conclude the same finding. Each test is calculated differently and addressed goodness of fit from a different perspective. In a perfect situation, all tests would give the same result. However, like with any other statistical decision where there are multiple tests to check the same assumption, we choose one test and use its result.

7.4 Association Statistics

The Association Statistics provide measures of association between the predicted probabilities and the observed responses of the ordered logistic regression model. This is a measure of the predictive capability of the model. Some common measures of association for categorical data, relevant for ordinal regression are, percent concordant, percent discordant, percent tied, pairs, Somers' D, Goodman- Kruskal Gamma, and Kendall's Tau-a.

7.4.1 Percent Concordant and Percent Discordant

Concordance as the name indicates is the measure of "agreement" between two items (or groups) using an attribute of interest. This term was often used in twin studies, to see if one twin had the attribute, if the other had it too. Then the twins were a concordant pair. In statistics, especially in non-parametric statistics it is used as a measure of association. In logistic regression it is used quite commonly. The idea is we look at a pair of observations (a,b), if observation a is lower than a observation b, and their expected values also follow that pattern (i.e., the expected value of a is lower than expected value of b), we call the pair concordant. If observed a is greater than b, but expected a is lower than b, we call the pair discordant. The percent concordance (PC) is looks at the percent of times the pairs are concordant looking at all possible pairs of observations. The percent discordant is similarly defined. Hence, higher PC is used as a diagnostic for good predictability of the model.

7.4.2 Somers' D

Somers' D was developed by Somers (1962) [73] and measures the proportion difference between the predicted probabilities and the observed responses including ties. This statistic ranges from -1 to 1, whereby a value of -1 indicates all pairs disagree, a value of 1 indicates that all pairs agree and a value of 0 indicates no association. That is, 1 indicates that as one variable goes up in value, the other also goes up in value; -1 indicates that as one variables goes up in value, the other variable goes down in value.

Somers' D is calculated by:

$$\text{Somers' D} = \frac{N_C - N_D}{N_C + N_D + N_T}$$

Where N_C is the number of pairs where the ranks of both variables agree, N_D is the number of pairs where the ranks of both variables disagree and N_T is the number of tied pairs.

Somer's D is usually used when the researcher wants to see the association between 2 ordinal variables. In this case, our 2 variables are the observed response (from our dataset) and the predicted response (based on our fitted model).

The R Code and result is below. Here, Somers' D is illustrated using the ordinal logit model (model_fit) from chapter 5.

```
library(DescTools) # # Loading the package DescTools

# Creating table with the observed and  predicted
    values
table = cbind(DATA$ConsensusEIPH, model_fit$fitted.
    values)

# Somers' D
SomersDelta(table)
```

In this analysis, we find a Somers' D of −0.003180452. It indicates no association between the predicted probabilities and observed responses because this value is very close to 0, even though it appears negative. This essentially shows that while the model was appropriate the predictive power of the model was poor.

7.4.3 Goodman-Kruskal Gamma

Gamma was developed by Goodman and Kruskal (1954) [30], (1963) [31] as to measure of association between the predicted probabilities and the observed responses. Like Somer's D, It ranges from −1 to 1, where 1 implies that all pairs agree, −1 implies that all pairs disagree and 0 implies no association.

Gamma is calculated by:

$$\gamma = \frac{N_C - N_D}{N_C + N_D}$$

Where N_C is the number of pairs where the ranks of both variables agree, and N_D is the number of pairs where the ranks of both variables disagree

The R Code and result is below. Here, γ is illustrated using the ordinal logit model (model_fit) from chapter 5.

```
library(DescTools) # # Loading the package DescTools

# Creating table with the observed and  predicted
    values
```

```
4 table = cbind(DATA$ConsensusEIPH, model_fit$fitted.
    values)
5
6 # Gamma
7 GoodmanKruskalGamma(table)
```

The value of Gamma for this model is -0.003183414. It indicates no association between the predicted probabilities and observed responses because this value is very close to 0, even though it appears negative. This agrees with Somer's D.

7.4.4 Kendall's Tau-A

Kendall's τ_A is named after Kendall (1938) [45], even though a similar approach in a time series context in 1897. τ_A is another measure of association between the predicted probabilities and the observed responses that ranges from -1 to 1. It measures the similarity of the variable orderings when they are ranked by each of the quantities. Like Somers' D and Gamma, 1 implies that all pairs agree, -1 implies that all pairs disagree, and 0 implies no association.

τ_A is calculated by:

$$\tau_a(C|R) = \frac{2N_C - 2N_D}{\frac{1}{2} \cdot n \cdot (n-1)}$$

Where n is the number of observations, N_C is the number of pairs where the ranks of both variables agree, and N_D is the number of pairs where the ranks of both variables disagree. As we see τ is closely related to PC defined above. The R Code and result is below. Here, τ_A is illustrated using the ordinal logit model (model_fit) from chapter 5.

```
1 library(DescTools) # # Loading the package DescTools
2
3 # Creating table with the observed and  predicted
    values
4 table = cbind(DATA$ConsensusEIPH, model_fit$fitted.
    values)
5
6 # Tau
7 KendallTauA(table)
```

The value -0.00211675 of Tau-a indicates no association between the predicted probabilities and observed responses because this value is very close to 0, even though it appears negative.

7.5 Model Fit Using Likelihood: AIC, BIC, and −2 Log L

A measure of goodness of fit that is a available in Maximum Likelihood Estimation is the criterion −2log(Likelihood). This essentially calculates the ML estimates and calculates this criterion using the estimated parameters.

$$-2log(\hat{L}), \hat{L} = L(Y, x, \hat{\theta}, M)$$

where, L represents the joint likelihood function of the model M, with observed data, Y and X and the ML estimates given by $\hat{\theta}$.

The idea is, if the model fits well, Likelihood will be larger. Log of the Likelihood is negative (as likelihood being a probability is less than 1). Hence, we take the negative of the log likelihood to get a positive number. Hence, our preferred $-2log(\hat{L})$ should be smaller for the most desirable model. Related statistic are AIC and BIC. AIC (Akaike Information Criterion) helps to make the trade-off between the model complexity and the goodness of fit based on the Likelihood equation. AIC penalizes for the number of parameters k that are estimated by the model.

$$AIC = -2log(\hat{L}) + k$$

A lower AIC indicates a better model fit.

BIC (Schwarz's Bayesian Information Criterion) is similar to AIC, but it places a higher penalty on model complexity.

$$BIC = -2log(\hat{L}) + klog(n)$$

Similar to AIC, a lower BIC indicates a better model fit.

The R Code and result is shown below. Here, AIC, BIC, and −2 Log L are illustrated using the ordinal logit model (model_fit) from chapter 5. To use these statistics, an additional model with only the intercept (no explanatory variables) should be fit.

```
1 # Fit an ordinal logit model with only the intercept
2 model_fit_intercept=polr(ordered(ConsensusEIPH) ~ 1,
    data=DATA, Hess=TRUE, method=c('logistic'))
3
4 # AIC
5 AIC(model_fit,model_fit_intercept)
6
7 # BIC
```

```
8  BIC(model_fit,model_fit_intercept)
9
10
11 library(lmtest) # # Loading the package lmtest for
       the # -2 Log L
12 lrtest(model_fit, model_fit_intercept)
```

Description: df [2 x 2]

	df <dbl>	AIC <dbl>
model_fit	9	2791.620
model_fit_intercept	4	2818.706

2 rows

(a) AIC

Description: df [2 x 2]

	df <dbl>	BIC <dbl>
model_fit	9	2836.407
model_fit_intercept	4	2838.612

2 rows

(b) BIC

```
Likelihood ratio test

Model 1: ordered(ConsensusEIPH) ~ Lasix + Surface + Finish + `Finish Time`
Model 2: ordered(ConsensusEIPH) ~ 1
  #Df  LogLik Df  Chisq Pr(>Chisq)
1   9 -1386.8
2   4 -1405.3 -5 37.086  5.757e-07 ***
---
Signif. codes:  0 '***' 0.001 '**' 0.01 '*' 0.05 '.' 0.1 ' ' 1
```

(c) -2 Log L

FIGURE 7.4
AIC, BIC, and −2 Log L

As mentioned earlier the metric −2 Log L (Log-likelihood) represents the overall goodness of fit of the model. It measures how well the model predicts the observed responses. R provides Log(likelihood) hence the values are negative. Our results for log (likelihood) are shown to be −1396 and −1405 for the model with predictors as opposed to model without predictors. Taking −2 of this indicates that the model with predictors fits slightly better. However, the numbers are fairly close in this case.

As seen with metric earlier the model with predictors has a lower AIC value (2791) compared to intercept only model (2818).

The same trend is true for BIC or SC value with for model with predictors(2836) being slightly lower than the intercept only model (2838). It suggests that the model with predictors fits better then intercept-only model. As our model is for illustration only, it reiterates some of the results seen in chapter 5, that the predictors do increase the fit of the model.

```
/* LACK OF FIT ORDINAL REG*/
PROC LOGISTIC DATA=dataset PLOTS=ALL;
MODEL Y=X1 X2 X3 X4 / LACKFIT;
run;
```

The LACKFIT option in PROC LOGISTIC calculates the Hosmer and Lemeshow test but not the other tests. Further, this will give us AIC, BIC, −2logL, Sommers D, Kendall's Tau, and the measures of concordance. No extra syntax is needed.

7.6 Proportional Odds Assumption

Before using the results of the proportional odds model, the proportional odds assumption must be checked. This assumes that each explanatory variable, X, has the same effect on each cumulative logit regardless of what the threshold values or categories. Since this assumption is fundamental to the design of the model, alternative methods should be used if this assumption is not met.

7.6.1 Brant-Wald Test

In 1990, Rollin Brant [14] proposed the Brant-Wald test to check for the proportionality assumption. This test checks to see if the estimated proportional odds model gives similar results as if a generalized ordinal logistic regression model is used. If the results are very similar, then the proportional odds assumption is not met because the generalized ordinal logistic model allows for estimates that are not proportional. To check for the difference, the Wald test is used.

For the Horse Dataset, the Brant-Wald test results are:
The R Code used is:

```
# Checking the Proportional Odds Assumption
library(brant) # Loading package to test the
    assumption
brant::brant(model_fit)
```

Test for	X2	df	probability
Omnibus	26.93	15	0.03
LasixY	3.78	3	0.29
SurfaceDirt	1.73	3	0.63
SurfaceTurf	0.54	3	0.91
Finish	2.97	3	0.4
FinishTime	1.46	3	0.69

H0: Parallel Regression Assumption holds

FIGURE 7.5
Brant-Wald test results

For this test, the Null Hypothesis is that the Proportional Odds Assumption holds. Hence, if a probability is less than your chosen α, we reject the null hypothesis and claim that the assumption does not hold. For this example, we use $\alpha = 0.05$. Since all the probabilities corresponding to the explanatory variables are greater than 0.05, we do not reject the Null Hypothesis. So, the proportional odds assumption does hold for this example. If at least one of the probabilities was less than α, we would reject the null hypothesis and claim that the proportional odds assumption is not met

If the proportional odds assumption does not hold, a partial proportional odds model or a generalized ordinal regression can be used.

7.6.2 Partial Proportional Odds Model

The partial proportional odds model was proposed by Peterson in 1990 [69], and is an extension of the proportional odds model but with the additional feature of allowing variables to not follow the proportional odds assumption. To fit the partial proportional odds model, extra parameters, γ_j, must be estimated that allows different clusters to deviate from the usual proportional odds cumulative logit model.

The partial proportional odds model is:

$$P(Y \geq j|X_i) = \frac{1}{1 + exp(\alpha_j - \beta^T X_i - \gamma_j^T T_i)}, \quad j = 1, \ldots, c-1$$

Assuming there are p explanatory variables, the additional parameter vector T_i consists of $q \leq p$ of the predictor variables on which the proportional odds assumption is to be tested or is not assumed and γ_j are the corresponding regression parameters. Again, the parameters α_j, β, and γ_j are estimated using maximum likelihood estimation [69]. When there are multiple categorical explanatory variables with multiple levels, there are many variables to estimate and the model becomes difficult to interpret.

To fit the Partial Proportional Odds Model in R, we use the following R Code:

Let's assume that the explanatory variables finish and surface met the parallel lines assumption. In the model, we need to specify that these two variables met the assumption.

```
1
2 library(VGAM)
3
4 # partial proportional odds model
5 model_fit_ppo <- vglm(ordered(ConsensusEIPH) ~ Lasix
    + Surface + Finish + 'Finish Time' ,
6              family=cumulative(link="logitlink",
7                                    parallel = TRUE ~ -1
    +
8                                      Surface + Finish),
9                                      data=DATA)
10 summary(model_fit_ppo)
```

```
1 /* ORDINAL LOGISTIC RUN*/
2 PROC LOGISTIC DATA=dataset PLOTS=ALL;
3 MODEL Y=X1 X2 X3 X4 / PCORR  CLODDS=both  CTABLE
    LACKFIT;
4 run;
```

The generic run of PROC LOGISTIC will give us the score test for the proportional odds assumption. No extra syntax is needed.

```
Call:
vglm(formula = ordered(ConsensusEIPH) ~ Lasix + Surface + Finish +
    `Finish Time`, family = cumulative(link = "logitlink", parallel =
TRUE ~
    -1 + Surface + Finish), data = DATA)

Coefficients:
                   Estimate Std. Error z value Pr(>|z|)
(Intercept):1      1.754529   0.583937   3.005 0.002659 **
(Intercept):2      3.503426   0.592660   5.911 3.39e-09 ***
(Intercept):3      5.316663   0.780011   6.816 9.35e-12 ***
(Intercept):4      5.644016   1.466813   3.848 0.000119 ***
LasixY:1           0.066100   0.210381   0.314 0.753377
LasixY:2           0.093175   0.230807   0.404 0.686440
LasixY:3           0.196073   0.422937   0.464 0.642935
LasixY:4          -0.912480   0.633361  -1.441 0.149671
SurfaceDirt       -1.455807   0.454524  -3.203 0.001360 **
SurfaceTurf       -1.399105   0.469673  -2.979 0.002893 **
Finish            -0.075583   0.020895  -3.617 0.000298 ***
`Finish Time`:1 -0.007853   0.004294  -1.829 0.067430 .
`Finish Time`:2 -0.010841   0.004319  -2.510 0.012081 *
`Finish Time`:3 -0.013030   0.007093  -1.837 0.066191 .
`Finish Time`:4  0.006481   0.016349   0.396 0.691818
---
Signif. codes:  0 '***' 0.001 '**' 0.01 '*' 0.05 '.' 0.1 ' ' 1
```

FIGURE 7.6
Partial proportional odds model results

```
Names of linear predictors: logitlink(P[Y<=1]), logitlink(P[Y<=2]),
logitlink(P[Y<=3]),
logitlink(P[Y<=4])

Residual deviance: 2762.166 on 4257 degrees of freedom

Log-likelihood: -1381.083 on 4257 degrees of freedom

Number of Fisher scoring iterations: 11

Warning: Hauck-Donner effect detected in the following estimate(s):
'(Intercept):3', '(Intercept):4'


Exponentiated coefficients:
       LasixY:1          LasixY:2          LasixY:3          LasixY:4
SurfaceDirt      SurfaceTurf
      1.0683331         1.0976534         1.2166157         0.4015272
0.2332120        0.2468177
         Finish `Finish Time`:1 `Finish Time`:2 `Finish Time`:3 `Finish
Time`:4
      0.9272027         0.9921777         0.9892177         0.9870543
1.0065016
```

R output with estimates for the partial proportional odds model

FIGURE 7.6
Partial proportional odds model results. *(Continued)*

From the results in Figure 7.6, we see that the variables that we assumed met the proportional odds assumptions had 1 estimate for each coefficient. However, the explanatory variables Lasix and finish time that we assumed did not meet the proportional odds assumption estimates for each intercept coefficient (that is, one for each cut-point). This means that the explanatory variables, Lasix, and finish time, has a different effect for each level of the outcome variable.

7.6.3 Other Models

Other models include the constrained partial proportional odds model discussed by Peterson in 1990 [69] and breaking up the ordinal logistic regression and using separate binary logistic regressions along with cumulative probabilities discussed by Bender in 1998 [8].

7.7 Autocorrelation

The assumption of uncorrelatedness is relevant in ordinal regression as much as general linear models. As a matter of fact, lack of this assumption is a difficult problem, as it is tricky to understand what it means to be correlated and how to measure correlation in an ordinal framework. With numerical variables, the concept of correlation is a tough one. If response is strictly categorical one is hard pressed to define correlation. So, with ordinal data where the issue of correlation is even trickier, we focus on association as we look at patterns in their frequency.

By definition, a categorical or ordinal variable can take a finite number of outcomes. Let us consider $\{Y_t\}_{t\in\mathbb{N}}$ to be an ordinal response that is measured over time. Numerical measures like correlation coefficient are not applicable for the qualitative range. It is known the for ordinal response, common measures like the Autocorrelation function (ACF) will change dramatically with the numerical scaling of the categories, hence, if we number categories 0, 1, .., 5, versus 0, 10,..., 50 our ACF could be different. In these cases it is apparent that ordinal response is not truly a number [10].

Just as in CDF formulation of the models, here too, one works with the probabilities of the response rather than the actual value of the response, [84].

Let us set up some notation, $\{Y_t\}_{t\in\mathbb{N}}$ is $\nu = \{0, ..., m\}$.

Define

$$p_j = P(Y_t = j), \mathbf{p} = (p_0, ..., p_m)' \quad and \quad s_k(\mathbf{p}) = \sum_{l=0}^{m} p_l^k \quad for \quad k \in \mathbb{N};$$

$$p_{ij}(k) = P(Y_t = i, Y_{t-k} = j) \quad and \quad p_{i|j}(k) = P(Y_t = i|Y_{t-k} = j).$$

To define autocorrelation we will define some measures. These are all related to the probability of the response being in a certain category at a time t and the probability at time $(t - k)$:

7.7.1 Serial Dependence Measures

Weiß and Göb (2013) [84] discussed several measures for serial dependence. Cramer's v is a measure for unsigned serial dependence.

$$v(k) = \sqrt{\frac{1}{m}\sum_{i,j=0}^{m}\frac{(p_{ij}(k) - p_i p_j)^2}{p_i p_j}}$$

Its value ranges from 0 to 1, where 0 indicates perfect serial independence at lag k, and 1 indicates perfect dependence at lag k. The sampling distribution of Cramer's v is asymptotically approximated by a χ^2 distribution.

Goodman and Kruskal's τ ((1954) [30]), given by

$$\tau(k) = \sqrt{\sum_{i,j=0}^{m}\frac{(p_{ij}(k) - p_i p_j)^2}{p_j(1 - s_2(\mathbf{p}))}},$$

is another measure for unsigned serial dependence. It is a number between 0 and 1, whereas 0 indicates perfect serial independence at lag k and 1 indicates perfect dependence at lag k.

To measure signed serial dependence, Cohen's κ is considered:

$$\kappa(k) = \frac{\sum_{j=0}^{m} p_{ij}(k) - s_2(\mathbf{p})}{1 - s_2(\mathbf{p})}, \quad or \quad \kappa^*(k) = \frac{1}{m}(\sum_{j=0}^{m} p_{i|j}(k) - 1).$$

The range of $\kappa(k)$ is $[-\frac{s_2(\mathbf{p})}{1-s_2(\mathbf{p})}, 1]$, where 0 represents serial independence with positive (negative) values indicating positive (negative) serial dependence at lag k. The distribution of the empirical Cohen's κ is asymptotically Normal distributed (Angel and Jose (2023) [52]).

We use our Horse Dataset to check for autocorrelation, to illustrate how to do this diagnostic. The R package *ctsfeatures* is often used to analyze categorical time series models. The R code and results are below:

```
library(ctsfeatures) # Loading library cstfeatures

# Cramers V

cramers_vi(as.factor(DATA$ConsensusEIPH),categories =
    factor(0:4),features = FALSE)

cramers_vi(as.factor(DATA$ConsensusEIPH),categories =
    factor(0:4),features = TRUE)

plot_cramers_vi(as.factor(DATA$ConsensusEIPH),
   categories = factor(0:4),max_lag = 3)



# Goodman Kruskal Tau

gk_tau(as.factor(DATA$ConsensusEIPH),categories =
   factor(0:4),features = FALSE)

gk_tau(as.factor(DATA$ConsensusEIPH),categories =
   factor(0:4),features = TRUE)



# Cohen Kappa

cohens_kappa(as.factor(DATA$ConsensusEIPH),categories
    = factor(0:4),features = FALSE)
```

```
24
25 cohens_kappa(as.factor(DATA$ConsensusEIPH),categories
      = factor(0:4),features = TRUE)
26
27 plot_cohens_kappa(as.factor(DATA$ConsensusEIPH),
     categories = factor(0:4), max_lag = 3)
```

SAS code for Autocorrelation
There is no direct measure for autocorrelation in Proc Logistic. However, we can trick SAS into finding the first order autocorrelation. We will show this when we look at chapter 10 and analyze the Heart data using SAS.

```
[1] 0.07224469
```

(a) result without features

```
              [,1]         [,2]         [,3]         [,4]         [,5]
[1,] 3.707976e-04 0.0004708729 1.421255e-08 1.116002e-03 0.0030995593
[2,] 6.845486e-06 0.0001911593 4.377620e-05 1.774045e-04 0.0040233564
[3,] 5.463883e-04 0.0018567516 1.612645e-04 9.990798e-05 0.0015834443
[4,] 1.288088e-05 0.0001220284 9.990798e-05 6.513359e-04 0.0036829934
[5,] 2.294325e-04 0.0002770505 1.857801e-03 5.165065e-08 0.0001961569
```

(b) result with features

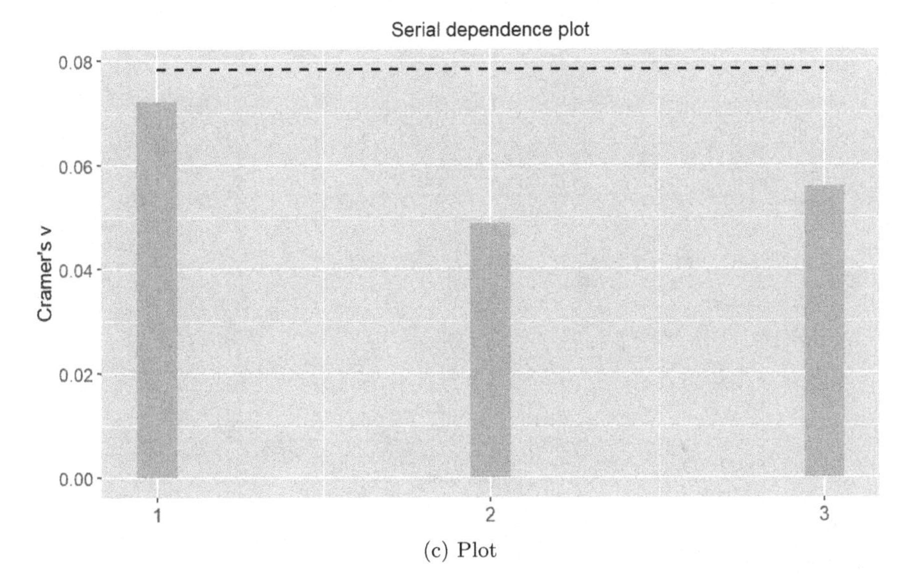

(c) Plot

FIGURE 7.7
Cramers V results

The autocorrelation from all three metrics is very small, close to zero. Example for Cohen's κ, is 0.07 without features, with features all around order of 10^{-4}.

```
[1] 0.006699031
```

(a) without features

```
             [,1]         [,2]         [,3]         [,4]         [,5]
[1,] 4.453369e-02 3.592139e-02 0.0265198966 0.0116931610 6.236353e-05
[2,] 4.253244e-02 3.960333e-02 0.0291376248 0.0068729802 6.236353e-03
[3,] 1.476278e-02 2.395757e-02 0.0104093193 0.0029232902 6.236353e-05
[4,] 1.472188e-03 1.818520e-03 0.0008661601 0.0001169316 5.612717e-04
[5,] 4.089411e-05 3.991266e-05 0.0001385856 0.0000129924 0.000000e+00
```

(b) with features

FIGURE 7.8
Goodman Kruskal τ

```
[1] -0.004335023
```

(a) result without features

```
[1]  0.0065805175 -0.0048410495 -0.0028812845 -0.0017157165 -0.0001961569
```

(b) result with features

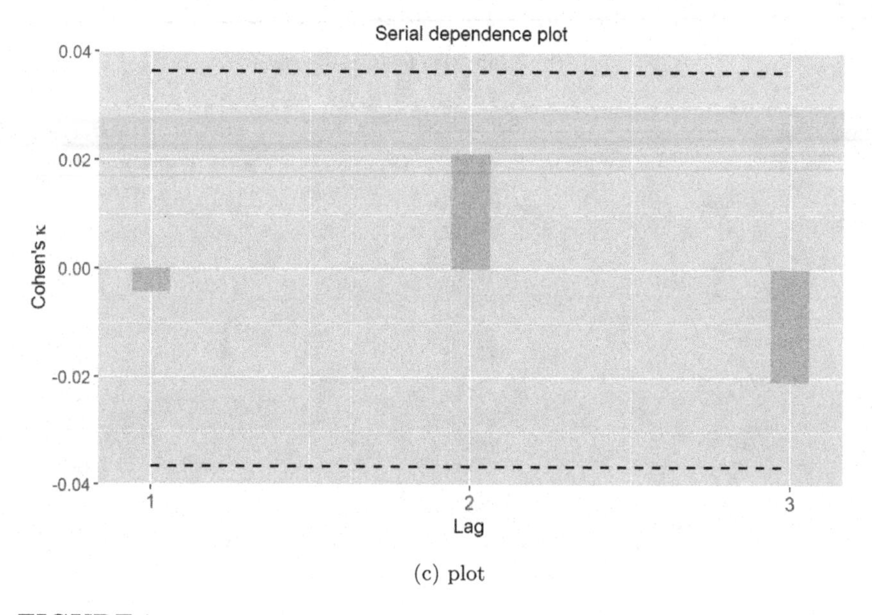

(c) plot

FIGURE 7.9
Cohen's κ results

For Cramer's V and Kendall's τ, the numbers are even smaller. This indicates that we are unable to detect autocorrelation in the Horse Dataset. This is not surprising as the horse data is not measured over time.

```
                GVIF Df GVIF^(1/(2*Df))
Lasix       1.123251  1        1.059835
Surface     1.096143  2        1.023215
Finish      1.011704  1        1.005835
FinishTime  1.122273  1        1.059374
```

FIGURE 7.10
VIF results

7.8 Multicollinearity

Multicollinearity occurs when two or more explanatory variables are highly correlated with each other. This makes the parameter estimates $\hat{\beta}$ unreliable and can artificially inflate the standard errors of the estimates. It is not only important to detect whether there is multicollinearity in the model, but also to know what degree of problem we have and determine which predictors cause the problem (Montgomery, Peck and Vining (2013) [58]).

7.8.1 Variance Inflation Factor

Variance Inflation Factor (VIF) is often used to check for Multicollinearity. VIF is calculated by:

$$VIF_i = \frac{1}{1 - R_i^2}$$

where R_i^2 is the un-adjusted coefficient of determination (R^2) when regressing the ith explanatory variable on the rest of the explanatory variables.

The R Code and result is below:

```
library(car)

# Variance Inflation Factor
car::vif(model_fit)
```

There is no direct measure for autocorrelation in Proc Logistic in SAS. However, we can trick SAS into finding the first-order autocorrelation. We will show this in chapter 10 when Heart data is analyzed using SAS.

Usually, a VIF value that is greater than 10 suggests that there is multicollinearity. This model does not have multicollinearity because all the GVIFs are less than 10.

7.8.2 Principal Component Regression

To improve the stability of the logistic model in the case of multicollinearity, principal component regression (PCLR), an extension of PCR, and partial least-square regression (PLS-LR) have been developed (Aguilera (2006) [3]).

PCR transforms a set of correlated variables into a set of linearly uncorrelated variables called principal components. The principal components are different linear combinations of the original predictor vectors. To find the principal components, we need the eigenvalues and eigenvectors of $\mathbf{X}'\mathbf{X}$. [58] Let $\Lambda = diag(\lambda_1, ..., \lambda_p)$, where $\lambda_1, ..., \lambda_p$ are the eigenvalues of $\mathbf{X}'\mathbf{X}$, and let $\mathbf{T}$ be the matrix whose columns are the corresponding eigenvectors. Then the linear model

$$\mathbf{y} = \mathbf{X}\beta + \epsilon$$

is equivalent to the principal component model

$$\mathbf{y} = \mathbf{Z}\alpha + \epsilon$$

where

$$\mathbf{Z} = \mathbf{XT}.\alpha = \mathbf{T}'\beta, \mathbf{T}'\mathbf{X}'\mathbf{XT} = \mathbf{Z}'\mathbf{Z} = \mathbf{\Lambda}.$$

The least squares estimate of $\hat{\alpha}$ is

$$\hat{\alpha} = (\mathbf{Z}'\mathbf{Z})^{-1}\mathbf{Z}'\mathbf{y}$$

and the covariance matrix of $\hat{\alpha}$ is

$$Var(\hat{\alpha}) = \sigma^2(\mathbf{Z}'\mathbf{Z})^{-1} = \sigma^2\Lambda^{-1}$$

Suppose that the principal component regressors are arranged in order of decreasing eigenvalues $\lambda_1 \geq \lambda_2 \geq ... \geq \lambda_p \geq 0$. Suppose further, that the last s eigenvalues are approximately equal to zero. In PCR, only the first $p - s$ principal component regressors are used to fit the model. The principal component parameter vector is

$$\hat{\beta}_{pc} = \begin{bmatrix} \hat{\alpha}_1 \\ \hat{\alpha}_2 \\ \vdots \\ \hat{\alpha}_{p-s} \\ 0 \\ \vdots \\ 0 \end{bmatrix}$$

The original parameter vector $\hat{\beta}_{pc}$ for principal component regression in terms of this principal component regressor is

$$\hat{\beta}_{pc} = \mathbf{T}\hat{\alpha}_{\mathbf{pc}}$$

By analogy, Escabias et al. [3] proposed the PCLR model by using a reduced set of pc's of the predictor variables as covariates of the logistic regression model. The main difference between PCR and PCLR is that in the later, the estimator $\hat{\alpha}_{pc}$in terms of the first s pc's is not the vector of the first s components of the estimator $\hat{\alpha}$ in terms of all the pc's.

PLS regression defines latent uncorrelated variables (PLS components) given by linear spans of the original predictors, and uses them as covariates of the regression model. [7] These linear spans take into account the relationship between the original explanatory variables and the response. Bastien et al. [7] adapted PLS to generalized linear models (PLS-GLR) and the particular case of logistic regression, where each one of the linear models is changed by the corresponding logit model meanwhile the linear fits are kept.

PLS-GLR of $\mathbf{y}$ on $\mathbf{x_1}, ..., \mathbf{x_p}$ with m components is written as:

$$g(\theta) = \sum_{h=1}^{m} c_h (\sum_{j=1}^{p} w_{hj}^{*} \mathbf{x}_j),$$

where θ is the mean of a continuous variable $\mathbf{y}$. The link function g is chosen by the user according to the probability distribution of $\mathbf{y}$ and and the model goodness of fit to the data. The new PLS components $\mathbf{t}_h = \sum_{j=1}^{p} w_{hj}^{*} \mathbf{x}_j$ are constrained to be orthogonal. The logarithm for computing a PLS-GLR model consists of three steps:

1. computation of the m PLS components $t_h (h = 1, ..., m)$;
2. generalized linear regression of $\mathbf{y}$ on the m retained PLS components;
3. formulation of the PLS-GLR regression model in terms of the original predictor variables

The interesting connection here is that these methods use similar ideas to those used in SEM, when the set of predictors are assumed to be correlated. There, CFA is done and the factor loadings as before are used. This again, emphasizes the gap of communication between the fields as the methods used, are theoretically similar, but called by different names.

It is important to note that the explanatory variables must be numerical to be used in algorithm in R. However, the Horse dataset did not have the issue of multicollinearity and so it does not make sense to we apply Principal Component Regression. For illustration purposes only, we provide the R Code using the explanatory variables finish, finish time, and Distance(meters) and the ordinal response variable Consensus EIPH. We do not present the output.

The R code and results are below:

```
library(plsRglm)  # loading library plsTglm

# Setting explanatory and response variables
y=ordered(DATA$ConsensusEIPH)
X=cbind(DATA$Finish,DATA$‘Finish Time‘,DATA$‘Distance
    (meters)‘)
colnames(X)=c('Finish','Finish_Time','Distance')

# Fitting Ordinal Logistic Regression using Principal
    Component Regression

modplsglm<-plsRglm(y, X, 1,model="pls-glm-polr")

# Model summary
summary(modplsglm)

# Coefficients from Model
coef.plsRglmmodel(modplsglm)
```

This is for illustration purpose only, as multicollinearity was not a problem for the Horse dataset.

7.9 Summary

In this chapter we talk about diagnostics and potential “fixes” when possible for ordinal regression. We talked about model “fitness” using Goodness of Fit, Association Tests, and Likelihood based criterion. If the model is not a good “fit”, the general solution is to look at other models. We looked at the proportional odds assumption and discussed some options when this assumption is not satisfied. We considered a way to look at potential autocorrelation. In general, ordinal regression with time series data is a much harder problem and we will not go into the details here, as that could warrant a text by itself. The research in the topic is still sparse and on-going. Finally we looked at the issue of multicollinearity and suggested methods as solutions should the problems arise. Diagnostics in ordinal regression is in general a much harder problem than the diagnostics in linear models. We have tried to provide an overview and code so that the practitioner can attempt to diagnose issues with ordinal data in a real life scenario. We used our data example to illustrate the issues.

8

Simulating Ordinal Models

8.1 Introduction

Simulations or computer experiments are common these days, and almost all statistical papers illustrate the proposed methods using simulations. However, the history of simulations or Monte Carlo simulations is traced back to World War II. The story goes that Polish-American physicist, Stanislaw Ulam, who was working in Los Alamos laboratory was sick. As a way of spending time, he played endless games of solitaire with cards and had little success. He was frustrated and wanted to figure out the probability of winning in a game of solitaire. As a card deck had 52 cards, he started working on the mathematical solution using combinatorics. It became so complicated that he decided there had to be another way. He then thought to play one hundred games of solitaire and count how many he won. He realized that would take a long time. At that point, his friend and colleague at Los Alamos, John Von Nuemann, had access to a ENIAC machine, or a state of the art computer of those times. They discussed the idea that it would be faster if they could program the computer to play 100 games and find the number of wins (Ekhard, 1987 [24]). They suggested this to Nicholas Metropolis (Metropolis and Ulam, 1949 [57]), another scientist at the laboratory. Together they came up with the idea of programming the computer to do experiments as a way to simulate real life and to find answers easier. The term "Monte Carlo" simulation is credited to Metropolis (Mazhdrakov et al, 2018 [53]). As the work they were doing was secret, they needed a code name for the process and Metropolis called it Monte Carlo based on the famous casinos and the fact that Ulman had an uncle who was a gambler at Monte Carlo. They then discussed that this process of getting the computer to perform a series of random experiments and then taking the results of these experiments as an "estimate" could work in other situations. Ulman and Von Nueman then used this idea to study the behavior of neurons around the time of World War II. The hydrogen bomb had connections with Monte Carlo simulations.

The general idea of Monte Carlo simulations is that we program the process for the computer and let the computer generate the process "randomly". Then, we take samples to study the process. The statistical principles behind the idea is the Law of Large Numbers. However, the key idea is making sure the

DOI: 10.1201/9781003020615-8

samples are taken randomly. In statistics, we often assume a certain distribution and then take random draws from that distribution. Hence, the harder statistical problem in reality is generating the "random" sample.

Most software have inbuilt functions for simulating data from some of the common distributions like Uniform, Normal, Bernoulli, etc. However, it is not a trivial process as the process has to be programmed correctly.

For researchers working on ordinal data, understanding the properties would require *simulating* ordinal data. The beauty of using simulations is that we know **the underlying truth is known**, as we know what structure and parameters we used to build the data. For example, if we perform the analysis using a proposed method with an existing dataset, we would have no idea how the method performed in general. Additionally, we would not be able to understand the variance of the estimators with a single dataset. This is where simulations are helpful. Given a population with known parameters, we could take many (often millions of samples) and look at the performance of the proposed method for each sample. In this chapter, we digress from real data analysis and talk about how to simulate ordinal data. We felt that this chapter was important as the common software do not provide functions to do this.

8.2 Simulating from a CDF Model

In this section, we describe a method to simulate data that has an ordinal response and multiple explanatory variables with the assumptions from the ordinal CDF models, with focus on the logit and probit models. So, we simulate a dataset that follows the structure:

$$F[P(Y \leq j|X)] = \alpha_j - \beta^T X, \qquad j = 1, \ldots, c-1 \tag{8.1}$$

where the explanatory (or predictor) parameters, β, determine how the estimates of the explanatory variables change when there are changes in the values of the explanatory variables, X, with the given structure in equation 8.1. The assumption about the structure of F determines whether we generate data under the logit (if F is assumed to be Logistic) or Probit (if F is assumed to be Normal).

To simulate reality, we make some assumptions about the parts of equation 8.1 to suit the population of interest. For example, we assume the probabilities in each of the "ordered" groups to help determine α_j. Additionally, we assume β_0 to be 0, as it is not estimable (being impossible to distinguish it from τ_j).

Under the assumption of $\beta_0 = 0$, the probability of the cut-points are directly related to the α_j parameter. We clarify the details as we explain the process.

In addition to assumptions about parts of equation 8.1, we must assume that:

1. All the observations are independent of each other.
2. Explanatory variables are independent of each other.
3. The proportional odds assumption, of equal slope across categories, holds.
4. There is an unobserved latent variable that is divided into groups (or chunks or sections) and the observed ordinal categories are manifestations of these groups.
5. The latent variable follows a **KNOWN** distribution, logistic for logit, Normal for Probit.
6. The group membership probabilities are known.

To discuss the process of simulating ordinal data with the ordinal logistic model and ordinal probit model structure, we discuss a simple dataset with one explanatory variable. The logic is exactly the same for multiple predictors.

8.3 Ordinal Probit Structure

To motivate the simulation of a simple dataset (one ordinal response and one explanatory variable), we will simulate the response variable, Y, consisting of $k = 5$ ordinal categories (0, 1, 2, 3, 4), and one continuous explanatory variable, x.

To do this, the structure of the ordinal probit model must be considered. Recall from chapter 5:

$$\Phi^{-1}[P_j] = \alpha_j - \beta^T X, \qquad j = 1, \ldots, c-1$$

where Φ is the CDF of the Normal distribution, and P_j is the cumulative probability, $P(Y \leq j|X)$, for a response Y with $c = 5$ categories.

Before we begin the simulations process, we must recall the relationship between the latent variable and the explanatory variables. After all, the structure of the model is based on a linear combination of the explanatory variable being related to the latent variable. From equation 5.7, for a single predictor x that relationship is:

$$\eta = \beta_0 + \beta_1 x + \epsilon$$

So, we see that our simulations procedure must first estimate the values of the latent variable based on the explanatory variables of interest. Once this link is made, then the latent variable η can be cut-up using thresholds to create the observed response. This action of cutting up the latent variable to create the observed response is defined by equation 5.6. Again to make matters simpler we are assuming β_0 as 0.

$$Y = \begin{cases} 1, & \text{if } \eta \leq \tau_1 \\ 2, & \text{if } \tau_1 < \eta \leq \tau_2 \\ \vdots & \vdots \\ c, & \text{if } \eta \geq \tau_{c-1} \end{cases}$$

This relationship where η has thresholds, $\tau_1, \tau_2, \tau_3, \tau_4$ allows Y, our ordinal response, to take values 0, 1, 2, 3 or 4. This is the key to creating our ordinal response groups.

The final step to this simulations process is the relationship between the probabilities of being observed in a certain group of the response, and the explanatory variables. In other words, do larger values of X indicate that Y will be more likely to be in a higher stage? For the ordinal probit model, this link function is the Standard Normal distribution and is described in equation 8.3 above.

To make this clear, we describe the following steps:

Step 1

Obtain estimates of the latent variable by using a linear combination of the explanatory variables and estimates of the cutpoints. Assumptions we make are:

1. We have the explanatory variables. Most often in the simulation process, we assume a structure or distribution for the explanatory variable. In real life, we assume that the X variables are non-stochastic, so no distribution is attached to the X. However in a simulation context we need to "get" the X somehow, and we generate it from a known distribution that mimics the structure we are trying to create. Common distributions for X are Normal, Uniform, Exponential depending on what we think the X represents.
2. We need to set the probabilities for the Y being in a particular category. For this, the proportions of observations in each stage that we would like to establish is assumed ahead of time. Keep in mind that these proportions should sum to one. Mathematically, we

need to define:

$$P(Y = \text{category } 1|X) = \pi_1$$
$$P(Y = \text{category } 2|X) = \pi_2$$
$$\dots$$
$$P(Y = \text{category c}|X) = \pi_c$$

From this, we can calculate the cumulative probabilities for the ordinal response. This gives the following general cumulative probabilities:

$$P(Y \leq \text{category } 1|X) = \pi_1$$
$$P(Y \leq \text{category } 2|X) = \pi_1 + \pi_2$$
$$\dots$$
$$P(Y \leq \text{category c}|X) = 1.0$$

3. We also need to set the value of the β, the slope (or slopes if we assume multiple predictors). This requires an understanding of how the estimates are interpreted (see sections 5.4). For the ordinal probit and logit model, β close to 0 implies a weak relationship between the explanatory variable and the ordinal response. Whereas positive slopes establish higher probabilities for certain stages and negative β represents higher probability in the reverse ordering.
4. Since this dataset is based on the ordinal probit model, we assume the error, ϵ is normally distributed.

   Hence, the four pieces we need to set in out simulations are:

   - Value of the X
   - The probabilities for each category π_j
   - The value and sign of β
   - The Normal distribution of ϵ, with $\mu = 0$ and a fixed σ.

   Using the equation 8.2, we can estimate α_j since we already know the observed values of the explanatory variables X, the estimates of the coefficients of those explanatory variables β_1, and the distribution of the error, ϵ. Rearranging the Ordinal Probit Model, we get :

$$\alpha_j = \Phi^{-1}[\pi_1 + \cdots + \pi_j] - x\beta_1, \qquad j = 1, \dots, c-1 \tag{8.2}$$

Step 2

In this second step, we use the functional relationship (Normal distribution cdf) along with the observed values of the explanatory variables, the assumed estimates of their coefficients, and the cut-points of the latent variable (as calculated in step 1), to estimate the cumulative probabilities for observing a given data point in the different response groups. Because we are using each individual data point, we get an estimate for each data point and each cumulative probability group.

We estimate the cumulative probabilities using the transformed equation 8.3 given below

$$\pi_1 + \cdots + \pi_j = \Phi(\alpha_j - x\beta_1,) \qquad j = 1, \ldots, c-1 \tag{8.3}$$

Step 3

This final step is to find the cumulative probabilities for each data-point to decide what category of the response that data point belongs to. The way this is done is by using a random variable that is uniformly distributed between 0 and 1 (so that each number is equally likely) to compare with the cumulative probabilities for the different categories. Using the uniform random variable simulates drawing a simple random variable as all the probabilities are equally likely. Further, the distribution of the CDF function is Uniformly distributed no matter what the original pdf is. Because the overall cumulative probability is one and the distance between the cumulative probabilities represent the estimated probability corresponding to each response outcome, the distributional assumptions of the model are preserved.

Pseudocode for the Ordinal Probit Structure

1. For each explanatory variable, generate n random deviates from the specified distribution.
2. Calculate each of the $j = c - 1$ intercepts.
3. For each data point, use the values of the explanatory variables, the calculated intercepts, and the assumed coefficients estimates to calculate the cumulative probabilities for the different response categories.
4. Assign response groups to each data point.

Data Simulation Example

Let's illustrate this simulation by simulating a dataset with a continuous numeric explanatory variable and an ordinal response variable with 5 categories.

Assumptions

1. Since we are using a continuous explanatory variable, we will assume that it follows a Normal distribution with mean of μ and

standard deviation of σ. In the code given we assumed mean of 83 and standard deviation of 16 (to match up with the variable, finish time from the Horse dataset).

2. We will also assume a symmetric distribution for the ordinal response variable. So, that is to say that we are simulating data with responses 0, 1, 2, 3, 4, each with equal probability. Then each group of the response will have 20% or (p=0.2) of the observations. That is:

$$P(Y = 0|X) = 0.2$$
$$P(Y = 1|X) = 0.2$$
$$P(Y = 2|X) = 0.2$$
$$P(Y = 3|X) = 0.2$$
$$P(Y = 4|X) = 0.2$$

Then the following cumulative probabilities apply:

$$P(Y \leq 0|X) = 0.2$$
$$P(Y \leq 1|X) = 0.4$$
$$P(Y \leq 2|X) = 0.6$$
$$P(Y \leq 3|X) = 0.8$$
$$P(Y \leq 4|X) = 1.0$$

3. Another assumption we make is about the size of the effect of the explanatory variable on the response. We set the β coefficient estimate to be 0.1.

4. We will assume that error term $\epsilon \sim N(0, 1)$.

These assumptions complete the latent variable equation as shown below:

$$\eta = \beta_0 + X_1\beta_1 + \epsilon$$

We will simulate a dataset of $n = 10$ observations with the above assumptions using the R Code below:

```
1 # This code simulates ordered data from the ordinal
      probit model
2
3 # Variables needed:
4 ## Sample Size/ number of observations
5
6 n = 10
7
```

```
8
9  ## Number of  explanatory variables
10
11 cov = 1
12
13
14 ## Number of response categories
15
16 resp = 5
17
18 # Objects to store parameters specific to the
      distribution that is needed for simulation
19
20 ## 1. vector to collect intercept values
21
22 alpha = vector(length=resp-1)
23
24
25 ## 2. Matrix to collect cumulative probabilities
26
27 q = matrix(0,n,resp-1)
28
29
30 ##################################
31 # AT THE END OF THIS SECTION, A DATAFRAME (v)
32 # IS CREATED THAT CONTAINS THE SIMULATED DATA - Y,
      X1
33 ###################################
34
35 ## Setting cumulative probabilities for the
      response
36
37 s = c(0.2,0.4,0.6,0.8,1)
38
39
40 ##  Vector that contains the estimates for the of
      the explanatory variables and the alpha vector.
       If there are more than 2 variables, list them
      separated by commas before the alpha vector
41
42 Beta = c(0.1, alpha)
43
44
45 ## Random deviates that are uniformly distributed
      between 0 and 1
```

```
46
47 unif=runif(n,0,1)
48
49
50 ## Creating dataframe with observed explanatory
     variables with a placeholder, Y, for the
     response. If there are additional explanatory
     variables, list them after "X1"
51
52 v = data.frame( X1 = rnorm(n,82,16), Y=0)
53 colnames(v)=c("X1","Y")
54
55
56 ## Calculating the intercepts. If there are
     additional explanatory variables, include them.
      These should be of the form: - Beta[2]*mean(v$
     X2) - ...
57
58 for (i in 1:(resp-1)) {
59   alpha[i]= qnorm(s[i]) - Beta[1]*mean(v$X1)
60   }
61
62
63 ## Calculating the cumulative probabilities. If
     there are additional explanatory variables,
     include them. These should be of the form: +
     Beta[2]*v$X2+ ...
64
65 for (i in 1:(resp-1)){
66   q[,i]=pnorm(alpha[i]+Beta[1]*v$X1)
67   }
68
69
70 ## Assigning Categories based on the cumulative
     probabilities
71
72 for (i in 1:n){
73   v[i,cov+1] = if (unif[i]<=q[i,1]) 1 else if (unif
     [i]<=q[i,2])  2 else if (unif[i]<=q[i,3])  3
     else if (unif[i]<=q[i,4])  4 else 5
74 }
75
76
77 print(v)   # viewing the simulated dataset
```

Here is the output from doing the simulation. The dataset v, with the explanatory variable, $X1$, and response variable, Y, looks like:

Description: df [10 × 2]

X1 <dbl>	Y <dbl>
72.35347	2
74.44534	4
71.83406	5
77.42762	2
84.20973	1
101.64209	1
69.17153	4
64.71372	5
79.47945	5
64.85184	3

1-10 of 10 rows

Simulated dataset following the ordinal probit structure

FIGURE 8.1
Simulated dataset with 10 observations, 5 response categories and one continuous explanatory variable

In this simulated dataset, we made general distributional assumptions about the explanatory and response variables for simplification. In a real situation, one would use theorized or known information about the variables to inform the simulations. For example, if considering the Horse data example, one would need to know some information about the distributions of the response, Consensus EIPH, and the explanatory variables, for example finish time. Here are the distributions from the Horse data example:

1. The distribution of counts for the ordinal response is right skewed.
2. The mean and standard deviation for the finish time variable are
3. The distribution of the finish time variable is bimodal.

If we wanted to generate a dataset with similar structure to the horse data (see Figure 8.1 below), it would make sense to choose a right skewed distribution of probabilities for the response variable (cumulative probabilities being: 0.34, 0.69, 0.91, 0.98, 1) and use random data from a bi-modal distribution (possibly a mixture of two Normal distributions) for the explanatory variable, with mean and standard deviation matching the real data.

A tibble: 5 × 2

ConsensusEIPH <fctr>	proportion <dbl>
0	0.34173669
1	0.35014006
2	0.22689076
3	0.06722689
4	0.01400560

5 rows

(a) Actual proportions of data in each response category from horse data

A tibble: 1 × 2

mean <dbl>	standard deviation <dbl>
82.96431	16.15879

1 row

(b) Actual mean and standard deviations explanatory variable

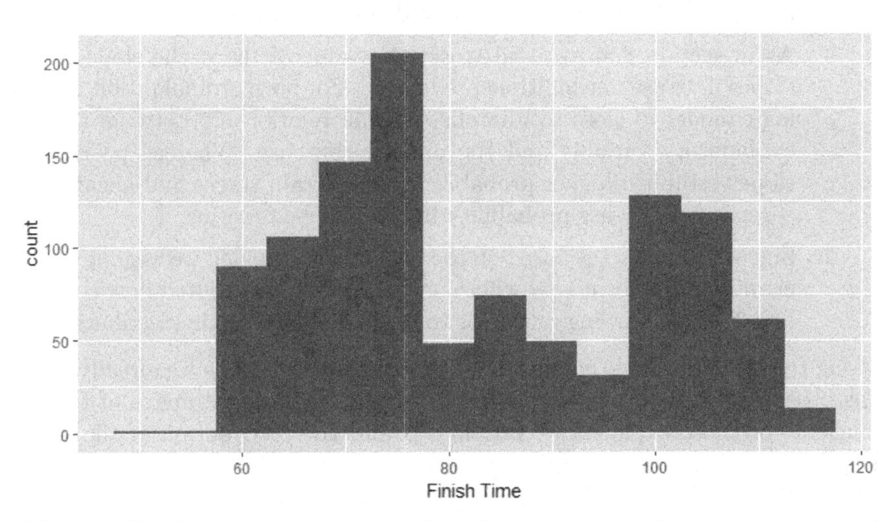

(c) Actual distribution of explanatory variable that suggests distribution is not unimodal

FIGURE 8.2
Summary statistics for and distribution of finish time

8.4 Ordinal Logistic Structure

The setup of the simulation method when the ordinal logistic structure is desired is based on distributional assumptions using the logistic distribution.

One could use a similar method as shown in the section 8.3 with simulating the ordinal probit model. However, we provide steps in this section for an alternative method. One could also use the method provided in this section to simulate from the ordinal probit model if appropriate distributional assumptions are made. The steps are:

Step 1

For each observation, estimate its location on the distribution of the latent variable η using the observed values of the explanatory variables, the estimates for the coefficients of the explanatory variables and the error. Assumptions we make are:

1. We have the explanatory variables, X. As is section 8.3, we assume a structure or distribution for the explanatory variable (while keeping in mind that in real life X variables are non-stochastic, so no distribution is attached to the X in real life). In a simulation context, we generate X from a known distribution that mimics the structure that we are trying to create.
2. We need to set the probabilities for the Y being in a particular category. For this, the proportions of observations in each stage that we would like to establish is assumed ahead of time. Keep in mind that these proportions should sum to one.
3. As in section 8.3, we need to set the value of the β the slope (or slopes if we assume multiple predictors). For both ordinal probit and logit model, β close to 0 implies a weak relationship between that explanatory variable and the ordinal response. Whereas positive slopes establish higher probabilities for certain stages and negative β represents higher probability in the reverse ordering.
4. Since this dataset is based on the ordinal logit model, we assume the error is of the form $\frac{U}{1-U}$ where u is uniformly distributed between 0 and 1, thus allowing our error to come from a logistic distribution.

Using the latent variable equation, we can estimate *eta* since we already know the observed values of the explanatory variables X, the estimates of the coefficients of those explanatory variables β and the distribution of the error, ϵ.

Step 2

In this step, we calculate the estimates of the cut-points of η. We use the cumulative probabilities based on the distribution of the response variable. Basically, the cut-points are defined by the quantiles at the cumulative probabilities. For this step, the assumptions we make are:

1. A logistic distribution of the response variable. For this, the proportions of observations in each outcome will suffice. Keep in mind that these proportions should sum to one. Mathematically, this and the associated cumulative probabilities are defined in the same way as in the ordinal probit structure.

Step 3

This final step is to assign response categories to each observation using the cut-points from step 2 and the location of each observation on the distribution of the latent variable from step 1. A response category is assigned if the estimates for the location on the distribution of η falls within the associated cut-points.

Pseudocode for Ordinal Logistic Structure

1. For each explanatory variable, generate n random deviates from the specified distribution.
2. Estimate the location of the n random deviates on the distribution of the latent variable.
3. Get the cut-points of the latent variable using cumulative probabilities.
4. Assign response groups to each data point.

Data Simulation Example

Let's illustrate this method by simulating a dataset with a continuous numeric explanatory variable and an ordinal response variable with 5 categories.

Assumptions

1. The continuous explanatory variable follows a Normal distribution with a known mean and standard deviation, μ and σ. The actual values and the distribution would depend upon the problem at hand.
2. In line 47 of the R code, we generate our X from a Normal distribution with $\mu = 82$ and $\sigma = 16$, illustrating a case with a fairly large standard deviation from Normal.
3. A symmetric distribution for the response, so that each group of the response will have 20% or (p=0.2) of the observations. That is:

$$P(Y = 0|X) = 0.2$$
$$P(Y = 1|X) = 0.2$$
$$P(Y = 2|X) = 0.2$$

$$P(Y = 3|X) = 0.2$$
$$P(Y = 4|X) = 0.2$$

Then the following cumulative probabilities apply:

$$P(Y \leq 0|X) = 0.2$$
$$P(Y \leq 1|X) = 0.4$$
$$P(Y \leq 2|X) = 0.6$$
$$P(Y \leq 3|X) = 0.8$$
$$P(Y \leq 4|X) = 1.0$$

4. The coefficient estimate is 0.1.
5. Error term $log\frac{u}{1-u}$ where u is uniformly distributed between 0 and 1.

We will simulate a dataset of $n = 10$ observations with the above assumptions using the R Code below:

```
# This code simulates ordered data from the ordinal
    logistic model

# Variables needed:
## Sample Size/ number of observations

n = 10


## Number of  explanatory variables

cov = 1


## Number of response categories

resp = 5

# Objects to store parameters specific to the
   distribution that is needed for simulation

## vector to collect intercept values

alpha = vector(length=resp-1)


```

```
25 #################################
26 # AT THE END OF THIS SECTION, A DATAFRAME (v)
27 # IS CREATED THAT CONTAINS THE SIMULATED DATA - Y,
     X1
28 ##################################
29
30 ## Setting cumulative probabilities for the
     response
31
32 s = c(0.2,0.4,0.6,0.8,1)
33
34
35 ##  Vector that contains the estimates for the of
     the explanatory variables and the alpha vector.
      If there are more than 2 variables, list them
     separated by commas before the alpha vector
36
37 Beta = c(0.2, alpha)
38
39
40 ## Random deviates from the uniform distribution
41
42 unif=runif(n,0,1)
43
44
45 ## Creating dataframe with observed explanatory
     variables with a placeholder, Y, for the
     response. If there are additional explanatory
     variables, list them after "X1"
46
47 v = data.frame( X1 = rnorm(n,82,16), Y=0)
48 colnames(v)=c("X1","Y")
49
50
51 ## Calculating the value of the observations on the
      latent variable distribution. If there are
     additional explanatory variables, include them.
      These should be of the form: - Beta[2]*v$X2 -
     ...
52
53 eta = 0 + Beta[1]*v$X1 +log(unif)-log(1-unif)
54
55 ## Calculating the cut-points. If there are more
     categories in the response variable, include
     them as additional cut-points.
```

```
56
57 tau1 = quantile(eta,s[1])
58 tau2 = quantile(eta,s[2])
59 tau3 = quantile(eta,s[3])
60 tau4 = quantile(eta,s[4])
61
62
63
64 ## Assigning Categories based on the cutpoints and
       the observation's value on latent variable
       distribution
65
66 Y = rep(NA,n)
67 Y[eta < tau1] = 1
68 Y[eta >= tau1 & eta < tau2] = 2
69 Y[eta >= tau2 & eta< tau3] = 3
70 Y[eta >= tau3 & eta < tau4] = 4
71 Y[eta >= tau4]<-5
72
73
74 ## Saving the response categories to the dataset.
75
76 v[,cov+1]<-data.frame(Y)
77
78 print(v)    # viewing the simulated dataset
```

Here is the output from doing the simulation. The dataset v, with explanatory variable, $X1$, and response variable, Y, looks like:

X1 <dbl>	Y <dbl>
66.84352	2
93.97023	3
80.12872	1
84.44252	1
117.03965	4
87.71178	3
125.46803	5
118.50323	5
87.18433	2
112.33707	4

1-10 of 10 rows

Simulated dataset following the ordinal logistic structure

FIGURE 8.3
Simulated dataset with 10 observations, 5 response categories and one continuous explanatory variable

8.5 Summary

In this chapter, we provide the logic and the code in R to simulate an ordinal response using numerical predictors. If we have multiple predictors, our η will incorporate that. Everything else will be similar. We wanted to put the logic and the code in this book, as we want the context of the problem to be clearly understood. Essentially, we provided slightly different methods for generating data from logit and probit. One could use either logic to generate data from logit or probit. In the ordinal logit structure simulation, we calculated the τ_j and then compared our observed values to the cut-points. In the Probit model, we used the cumulative probabilities to assign the group memberships. We provided the code in R. But what we are providing is the logic and one could use any language to implement the logic.

Part V

Overall Summary

9

Data Story: Analysis of the Heart Data

All through the book we have used the Horse Example to illustrate how to analyze a dataset with ordinal data. We have provided code and output from R and snippets of generic code from SAS. In this chapter we will focus on the Heart example and do the analysis from beginning to end. To illustrate, we will use SAS and the output from SAS. However, we will provide the relevant snippets of generic code from R as well. Our hope, with both examples completely worked out with code and output, is that readers can follow along with the statistics and how they are applied to real data.

9.1 Introduction

As mentioned in chapter 1, this dataset is part of a much larger study that has been analyzed using different response variables, Hoekstra et al (2014) [34], Hatch, (2003) [33], Dasgupta et al, (2000) [66]. Many response variables were collected along with a multitude of explanatory variables. For this illustration we will use coronary artery calcification stage (0, 1, 2, 3 or 3), or "risk stage" as the response variable. We will refer to the response stages as "risk". The explanatory variables that we explore are glucose, gender, body mass index (BMI), total cholesterol, cholesterol ratio (total cholesterol/LDL), red blood cells (RBC), white blood cells (WBC), potassium, diastolic blood pressure, systolic blood pressure, lactate dehydrogenase (LDH), and mean corpuscular hemoglobin (MCH). The dataset has 691 participants and the goal is to try to understand which of the twelve predictors contributed to a higher stage of coronary calcification. The explanatory variables were just a few that we selected from the list.

9.2 Exploratory Data Analysis

For the exploratory data analysis stage, we would like to understand the nuances of the dataset using graphical and numerical summaries. Since our response is ordinal we construct barplots, pie charts, or frequency tables to understand the shape and structure of the response. Using our the Heart

DOI: 10.1201/9781003020615-9

Dataset with the response variable risk, the relevant SAS code for pie chart and barplot:

```
PROC GCHART data=heart;
PIE risk;
HBAR risk;
run;
```

And the resulting output is:

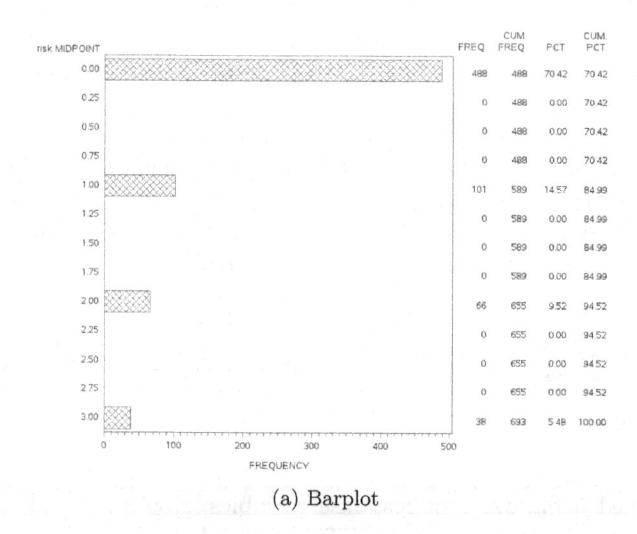

(a) Barplot

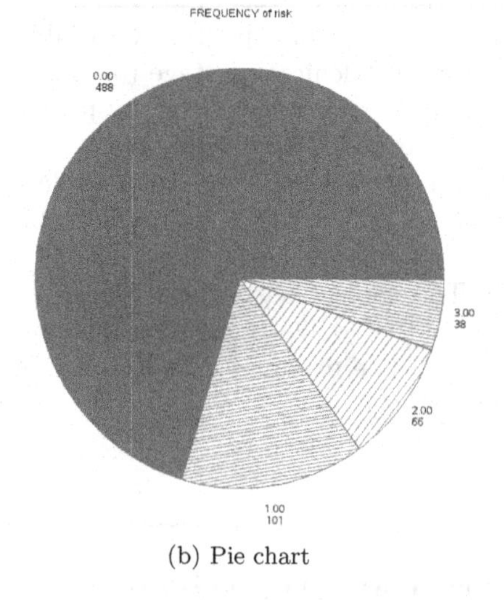

(b) Pie chart

FIGURE 9.1
Exploratory graphs

From the graphs in Figure 9.1, it is evident that most of the participants had a risk stage of 0, 488 out of 693(70%). This was followed by 101 who had a risk stage of 1: 14.57%, with 66 or 9.52% in risk stage 2, and 38 or 5.48% as the highest stage of risk at 3. This makes sense, as this group of participants were primarily healthy young people without any outward sign of disease.

To understand the structure of the explanatory variables (as they are all numerical) we use a matrix plot. A matrix plot allows us to plot all the variables as a matrix of scatter plots. This allows us to get an idea of how the response and explanatory variables are related to each other. While the plots are small as the number of predictors is large, it is still a quick and dirty overview of the data.

The relevant SAS code is:

```
PROC SGSCATTER DATA=heart;
MATRIX risk glucose sex BMI cholesterol cholratio RBC
    WBC potassium diastolic systolic LDH MCH ;
run;
```

Several options can be added to make the plots look nicer, but this code is the barebones of it.

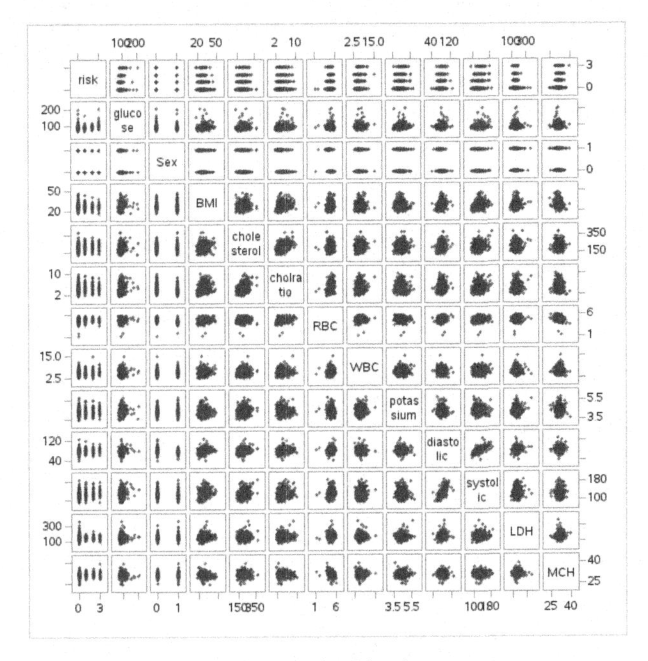

FIGURE 9.2
Scatterplot matrix

The scatter plot matrix in Figure 9.2 is given. The first row of plots indicates the relationship between the response, risk, and the twelve predictors. The scatter plots are comprised of five rows (indicating that our response is in five groups). If we look at the plot of sex and risk (third square on the top row) it looks like it is comprised of ten clusters of points. Indicating a ordinal response (5 groups) and a predictor with two categories. Scatter plots are hard to read with binary or ordinal response, so we would need to look at LOESS plots for some of the relevant variables like we did in section 2.2.4. We provide LOESS plots separately because interpreting LOESS lines in the matrix becomes more difficult as the number of predictors increase. The scatter plot matrix gives us more of an idea of relationships among the predictors. From the scatterplot matrix, it appears that there is a linear relationship between the pairs (1) total cholesterol and cholesterol ratio, and (2) diastolic blood pressure and systolic blood pressure. Given our knowledge of these predictors, they make sense.

To look at the strength of the relationship we can look at the correlations as well. To find correlation the SAS Code is given below:

```
PROC CORR data=heart;
VAR risk glucose sex BMI cholesterol cholratio
RBC WBC potassium diastolic systolic LDH MCH ;
run;
```

PROC CORR produces the following output:

The CORR Procedure

13 Variables:	risk glucose Sex BMI cholesterol cholratio RBC WBC potassium diastolic systolic LDH MCH

Simple Statistics							
Variable	N	Mean	Std Dev	Sum	Minimum	Maximum	Label
risk	691	0.50072	0.87869	346.00000	0	3.00000	
glucose	690	92.02609	12.46100	63498	68.00000	209.00000	glucose
Sex	691	0.28365	0.45109	196.00000	0	1.00000	Sex
BMI	687	27.22247	4.41834	18702	16.59300	47.07654	BMI
cholesterol	690	210.66667	37.51856	145360	126.00000	380.00000	cholesterol
cholratio	689	4.65689	1.47850	3209	1.70000	10.20000	cholratio
RBC	687	4.81461	0.44507	3308	1.29000	5.85000	RBC
WBC	688	5.82006	1.42149	4004	2.40000	15.00000	WBC
potassium	690	4.33768	0.33085	2993	3.50000	5.80000	potassium
diastolic	633	80.90679	9.64705	51214	42.00000	130.00000	diastolic
systolic	633	126.66667	15.39330	80180	90.00000	190.00000	systolic
LDH	690	154.68986	28.88332	106736	71.00000	369.00000	LDH
MCH	687	31.18064	1.55697	21421	24.00000	38.90000	MCH

(a) Simple Statistics

FIGURE 9.3
Correlation and simple statistics tables

Pearson Correlation Coefficients
Prob > |r| under H0: Rho=0
Number of Observations

	risk	glucose	Sex	BMI	cholesterol	cholratio	RBC	WBC	potassium	diastolic	systolic	LDH	MCH
risk	1.00000 691	0.19701 <.0001 690	-0.17602 <.0001 691	0.00139 0.9710 687	0.14505 0.0001 690	0.10352 0.0065 689	0.06864 0.0722 687	0.09622 0.0116 688	0.13903 0.0002 690	0.16881 <.0001 633	0.17314 <.0001 633	-0.06109 0.1089 690	0.16260 <.0001 687
glucose **glucose**	0.19701 <.0001 690	1.00000 690	-0.10497 0.0058 690	0.21698 <.0001 686	0.14868 <.0001 690	0.21208 <.0001 689	0.09577 0.0121 686	0.08453 0.0267 687	0.10868 0.0043 690	0.16567 <.0001 633	0.21532 <.0001 633	0.02495 0.5130 690	-0.04799 0.2093 686
Sex **Sex**	-0.17602 <.0001 691	-0.10497 0.0058 690	1.00000 691	-0.05140 0.1784 687	-0.03305 0.3860 690	-0.31816 <.0001 689	-0.52515 <.0001 687	0.09061 0.0174 688	0.03556 0.3510 690	-0.26323 <.0001 633	-0.14519 0.0002 633	-0.03363 0.3778 690	0.00455 0.9053 687
BMI **BMI**	0.00139 0.9710 687	0.21698 <.0001 686	-0.05140 0.1784 687	1.00000 687	0.18771 <.0001 686	0.32754 <.0001 685	0.16797 <.0001 683	0.14685 0.0001 684	-0.03080 0.4206 686	0.29183 <.0001 630	0.28949 <.0001 630	0.14788 0.0001 686	-0.15244 <.0001 683
cholesterol **cholesterol**	0.14505 0.0001 690	0.14868 <.0001 690	-0.03305 0.3860 690	0.18771 <.0001 686	1.00000 690	0.47542 <.0001 689	0.06726 0.0783 686	0.06093 0.1106 687	0.06726 0.0775 690	0.17771 <.0001 633	0.22115 <.0001 633	0.11875 0.0018 690	0.03130 0.4131 686
cholratio **cholratio**	0.10352 0.0065 689	0.21208 <.0001 689	-0.31816 <.0001 689	0.32754 <.0001 685	0.47542 <.0001 689	1.00000 689	0.33453 <.0001 685	0.09649 0.0115 686	-0.02448 0.5212 689	0.21532 <.0001 633	0.18580 <.0001 633	0.07986 0.0361 689	-0.07798 0.0413 685
RBC **RBC**	0.06864 0.0722 687	0.09577 0.0121 686	-0.52515 <.0001 687	0.16797 <.0001 683	0.06726 0.0783 686	0.33453 <.0001 685	1.00000 687	0.04792 0.2097 687	-0.06330 0.0976 686	0.20623 <.0001 629	0.10655 0.0075 629	0.03931 0.3039 686	-0.38006 <.0001 687
WBC **WBC**	0.09622 0.0116 688	0.08453 0.0267 687	0.09061 0.0174 688	0.14685 0.0001 684	0.06093 0.1106 687	0.09649 0.0115 686	0.04792 0.2097 687	1.00000 688	0.06289 0.0996 687	0.01875 0.6385 630	0.10066 0.0115 630	0.04184 0.2734 687	-0.03443 0.3676 687
potassium **potassium**	0.13903 0.0002 690	0.10868 0.0043 690	0.03556 0.3510 690	-0.03080 0.4206 686	0.06726 0.0775 690	-0.02448 0.5212 689	-0.06330 0.0976 686	0.06289 0.0996 687	1.00000 690	0.00911 0.8190 633	-0.04487 0.2596 633	0.01456 0.7026 690	0.04912 0.1989 686
diastolic **diastolic**	0.16881 <.0001 633	0.16567 <.0001 633	-0.26323 <.0001 633	0.29183 <.0001 630	0.17771 <.0001 633	0.21532 <.0001 633	0.20623 <.0001 629	0.01875 0.6385 630	0.00911 0.8190 633	1.00000 633	0.65402 <.0001 633	0.11108 0.0051 633	0.00491 0.9022 629
systolic **systolic**	0.17314 <.0001 633	0.21532 <.0001 633	-0.14519 0.0002 633	0.28949 <.0001 630	0.22115 <.0001 633	0.18580 <.0001 633	0.10655 0.0075 629	0.10066 0.0115 630	-0.04487 0.2596 633	0.65402 <.0001 633	1.00000 633	0.09697 0.0147 633	0.02272 0.5696 629
LDH **LDH**	-0.06109 0.1089 690	0.02495 0.5130 690	-0.03363 0.3778 690	0.14788 0.0001 686	0.11875 0.0018 690	0.07986 0.0361 689	0.03931 0.3039 686	0.04184 0.2734 687	0.01456 0.7026 690	0.11108 0.0051 633	0.09697 0.0147 633	1.00000 690	0.05301 0.1655 686
MCH **MCH**	0.16260 <.0001 687	-0.04799 0.2093 686	0.00455 0.9053 687	-0.15244 <.0001 683	0.03130 0.4131 686	-0.07798 0.0413 685	-0.38006 <.0001 687	-0.03443 0.3676 687	0.04912 0.1989 686	0.00491 0.9022 629	0.02272 0.5696 629	0.05301 0.1655 686	1.00000 687

(b) Correlation Table

FIGURE 9.3
Correlation and simple statistics tables. *(Continued)*

Based on Figure 9.3 (b), the correlations between the explanatory variables and the response risk is not high (but with ordinal responses we are not expecting high). To understand this, we need to realize no matter what the value of the predictor is, the response can only take values from 1 to 5. This limited range of Y makes correlations between ordinal and other variables on the smaller side. The correlations in decreasing order to risk are for the variables: glucose, sex, systolic, diastolic, MCH, cholesterol, and potassium. The others, BMI, cholratio, RBC, WBC, and LDH show less relationship with risk. Among the predictors, systolic and diastolic show a correlation of 0.6 and cholesterol and cholratio show a correlation of 0.4. SAS also provides the simple statistics in Figure 9.3 (a) for the predictors in terms of mean, standard deviation, minimum, and maximum are also provided.

To understand how the explanatory variables and response variable are related (especially for ordinal or binary response), we can do scatterplots with a LOESS line as was suggested for looking at relationships between the ordinal response variable and the explanatory variables. In this case, we will illustrate the plots for risk and glucose, risk and sex and risk and BMI as they had the strongest relationships or a non-linear relationship. The SAS code is given below:

```
PROC SGPLOT data=heart ;
LOESS x=glucose y=risk/ smooth=.5;
run;
```

PROC SGPLOT gives the following output:

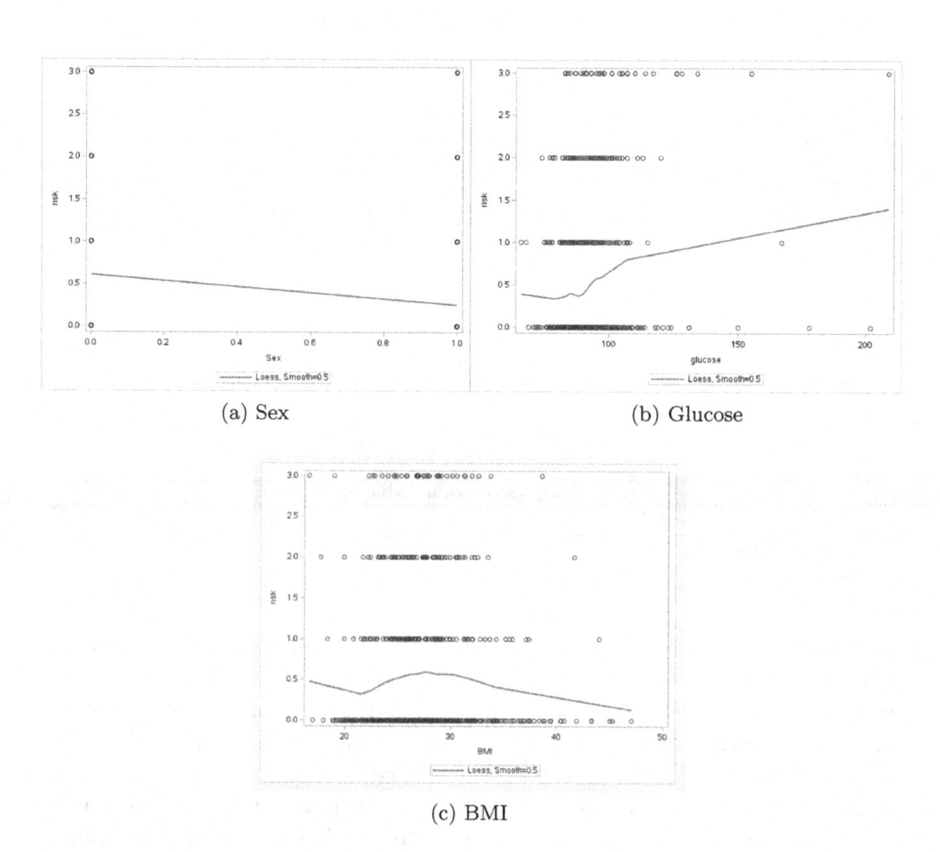

(a) Sex

(b) Glucose

(c) BMI

FIGURE 9.4
Loess Plots

For glucose we see a positive relationship and for sex we see a negative relationship. For BMI (where the correlation was very weak) we see somewhat of a quadratic relationship. These plots are very helpful for understanding patterns.

The other scatterplots with LOESS curves for the relationship with the predictors and the response predictors are provided below:

Our findings from EDA can be summarized as:

- The ordinal response for risk is skewed with most of the people in the stage showing no risk.
- Several predictors appear to be related to the response risk, some positively and some negatively.

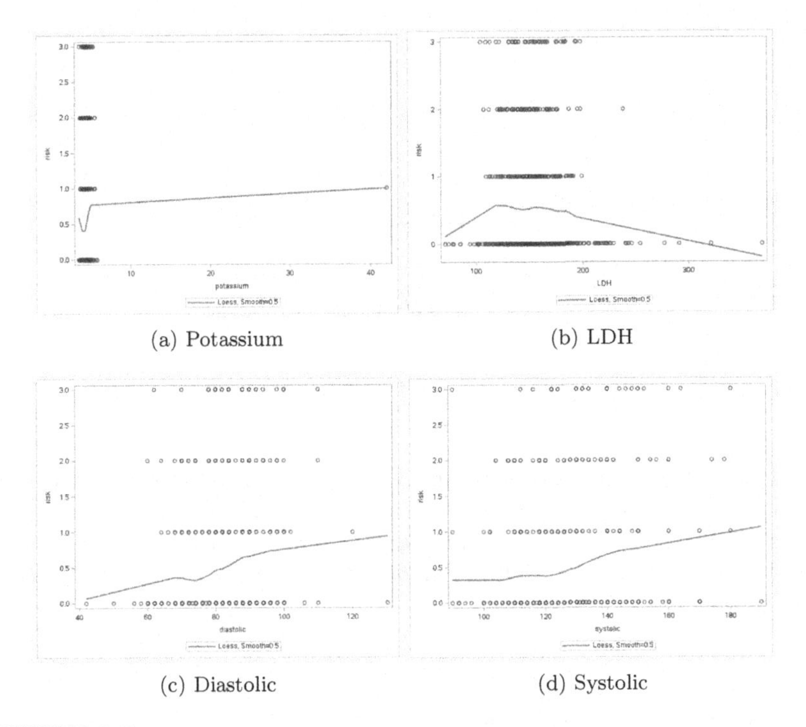

(a) Potassium (b) LDH

(c) Diastolic (d) Systolic

FIGURE 9.5
Loess plots

9.3 Analysis Using Ordinal Logistic Model

The Heart data has twelve explanatory variables, and our scatter plots and correlation indicate that not all these predictors explain the stage of risk. Nevertheless, we run the ordinal logit model and ordinal probit model with

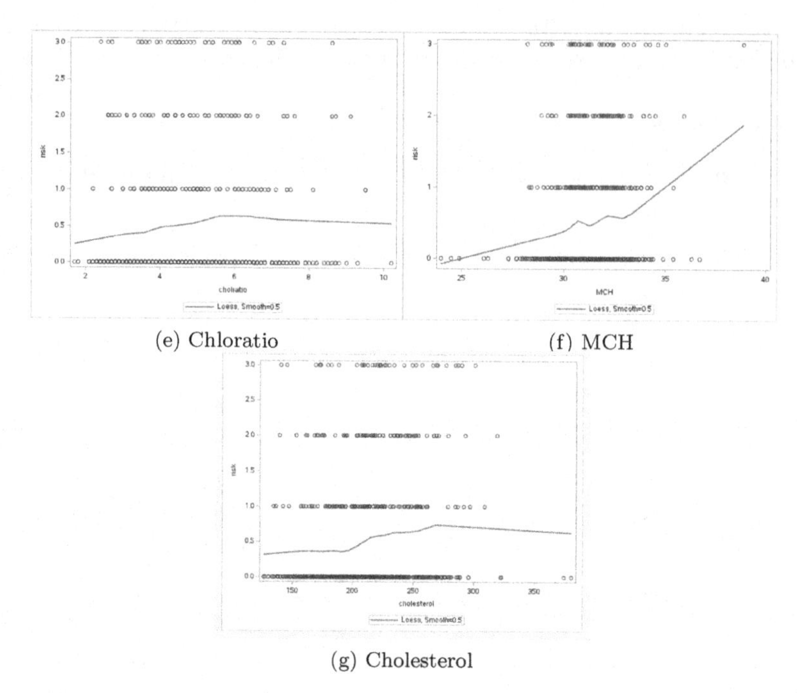

(e) Chloratio (f) MCH

(g) Cholesterol

FIGURE 9.5
Loess plots. *(Continued)*

all predictors in it. In regression, this is called running the FULL model. As the number of predictors is not too large, this is possible and preferable, even if this is not the model we finally decide as our "final model". We refer to your final model as the "CHOSEN model". The SAS and output and code are given below:

```
/* FULL Model*/
PROC LOGISTIC DATA=heart DESCENDING;
MODEL risk=glucose sex BMI cholesterol cholratio
RBC WBC potassium diastolic systolic LDH MCH ;
run;
```

When the ordinal response is presented using numbers and the order of the number correctly matches the increasing or decreasing trend of the variable, you can use the DESCENDING option in SAS to ensure that the lowest group is taken as the baseline group.

Model Convergence Status
Convergence criterion (GCONV=1E-8) satisfied.

Score Test for the Proportional Odds Assumption		
Chi-Square	DF	Pr > ChiSq
45.1864	24	0.0055

Model Fit Statistics		
Criterion	Intercept Only	Intercept and Covariates
AIC	1145.340	1076.076
SC	1158.667	1142.714
-2 Log L	1139.340	1046.076

Testing Global Null Hypothesis: BETA=0			
Test	Chi-Square	DF	Pr > ChiSq
Likelihood Ratio	93.2640	12	<.0001
Score	81.6588	12	<.0001
Wald	78.8927	12	<.0001

(a) Model convergence and fit

Analysis of Maximum Likelihood Estimates						
Parameter		DF	Estimate	Standard Error	Wald Chi-Square	Pr > ChiSq
Intercept	3	1	-18.6129	3.4243	29.5450	<.0001
Intercept	2	1	-17.2938	3.4126	25.6815	<.0001
Intercept	1	1	-16.2641	3.4033	22.8380	<.0001
glucose		1	0.0255	0.00737	11.9477	0.0005
Sex		1	-1.0700	0.3016	12.5864	0.0004
BMI		1	-0.0207	0.0247	0.6988	0.4032
cholesterol		1	0.00847	0.00291	8.4578	0.0036
cholratio		1	-0.0481	0.0780	0.3804	0.5374
RBC		1	0.3781	0.2915	1.6824	0.1946
WBC		1	0.0288	0.0304	0.8981	0.3433
potassium		1	0.0637	0.0474	1.8069	0.1789
diastolic		1	0.0152	0.0126	1.4642	0.2263
systolic		1	0.00919	0.00783	1.3765	0.2407
LDH		1	-0.0101	0.00362	7.8451	0.0051
MCH		1	0.2919	0.0733	15.8714	<.0001

(b) Analysis of MLE

FIGURE 9.6
Full model output

Odds Ratio Estimates			
Effect	Point Estimate	95% Wald Confidence Limits	
glucose	1.026	1.011	1.041
Sex	0.343	0.190	0.619
BMI	0.980	0.933	1.028
cholesterol	1.009	1.003	1.014
cholratio	0.953	0.818	1.110
RBC	1.460	0.824	2.585
WBC	1.029	0.970	1.092
potassium	1.066	0.971	1.170
diastolic	1.015	0.991	1.041
systolic	1.009	0.994	1.025
LDH	0.990	0.983	0.997
MCH	1.339	1.160	1.546

(c) OR estimates

Association of Predicted Probabilities and Observed Responses			
Percent Concordant	70.1	Somers' D	0.407
Percent Discordant	29.4	Gamma	0.409
Percent Tied	0.5	Tau-a	0.192
Pairs	93033	c	0.704

(d) Association Statistics

FIGURE 9.6
Full model output. *(Continued)*

The model results in Figure 9.6 are given with its convergence status, the frequency, the Maximum Likelihood estimates, the standard errors, and p-values. confidence interval for log odds and association measure is also provided by default in SAS. The model has a concordance of 70%, several predictors show p-values less than 0.10 and the model converged. However, we see that diastolic and systolic blood pressure that came up somewhat positively correlated with risk had some of the larger p-values. This often tells us that we should try and see if a parsimonious model would be more appropriate. As a result, we will look at model selection.

9.4 Model Selection

In chapter 7, we discussed model selection briefly. The basic idea is to choose the best model based on model fit. This can be done using model selection techniques like stepwise model selection, where many models are fit using subsets of the predictor variables and their model fit statistics, association statistics, AIC, BIC, −2Log L values are used to compare between models. Then, the best model is chosen as the model that had the best criteria based on these diagnostics. For the Heart data example, we will use the most common stepwise model selection and SAS' default diagnostic to choose the best model. The SAS code is provided below:

```
/*MODEL SELECTION*/
PROC LOGISTIC DATA=heart DESCENDING;
MODEL risk=glucose sex BMI cholesterol cholratio
RBC WBC potassium diastolic systolic LDH MCH /
SELECTION =STEPWISE SLENTRY=.25 SLSTAY=.25;
run;
```

For space considerations we provide Step 1, Step 9 (the final step), and the final model of the Stepwise selection output.

Step 1. Effect Sex entered:

Model Convergence Status
Convergence criterion (GCONV=1E-8) satisfied.

Score Test for the Proportional Odds Assumption		
Chi-Square	DF	Pr > ChiSq
0.1060	2	0.9484

Model Fit Statistics		
Criterion	Intercept Only	Intercept and Covariates
AIC	1145.340	1116.888
SC	1158.667	1134.658
-2 Log L	1139.340	1108.888

Testing Global Null Hypothesis: BETA=0			
Test	Chi-Square	DF	Pr > ChiSq
Likelihood Ratio	30.4515	1	<.0001
Score	26.9829	1	<.0001
Wald	25.1959	1	<.0001

Residual Chi-Square Test		
Chi-Square	DF	Pr > ChiSq
59.4565	11	<.0001

Note: No effects for the model in Step 1 are removed.

(a) Step 1

FIGURE 9.7
Stepwise selection output for Step 1 and Step 9

Our results from Figures 9.7 and 9.8 show that stepwise for model selection used nine steps to select the best model. In each step, one new variable was added to or removed from the model and the diagnostics were checked. The results indicate that 7 of the twelve predictors came up as relevant in the model. We used SLENTRY (Significance level for entry in the model), SLSTAY (Significance level of staying in the model) of a somewhat liberal 0.25. This means

Step 9. Effect systolic is removed:

Model Convergence Status
Convergence criterion (GCONV=1E-8) satisfied.

Score Test for the Proportional Odds Assumption		
Chi-Square	DF	Pr > ChiSq
29.9354	14	0.0078

Model Fit Statistics		
Criterion	Intercept Only	Intercept and Covariates
AIC	1145.340	1070.819
SC	1158.667	1115.244
-2 Log L	1139.340	1050.819

Testing Global Null Hypothesis: BETA=0			
Test	Chi-Square	DF	Pr > ChiSq
Likelihood Ratio	88.5208	7	<.0001
Score	77.4241	7	<.0001
Wald	76.3939	7	<.0001

Residual Chi-Square Test		
Chi-Square	DF	Pr > ChiSq
4.6716	5	0.4573

Note: No effects for the model in Step 9 are removed.

Note: Model building terminates because the last effect entered is removed by the Wald statistic criterion.

(b) Step 9

FIGURE 9.7
Stepwise selection output for Step 1 and Step 9. *(Continued)*

we allow predictors with p-values up to 0.25 to enter the model and predictors with p-values up to 0.25 to stay in the model. As this is model selection, we would like to have as many predictors to have a chance of entering the model.

The steps included the variables in the following order for the first eight steps: Sex, MCH, Glucose, Cholesterol, LDH, Diastolic, Potassium, and Systolic. There was a ninth step that removed Systolic from the model and provided a seven variable model as the CHOSEN model. We see that of the seven predictors relevant in the model glucose, MCH, diastolic, and cholesterol are all positively related with p-values less than .01. Potassium has a p-value around 0.17. Sex and LDH appear to be negatively related to risk stage. This chosen model had a percent concordance (PC) of 69.8% compared to the FULL model of 70.1%. However, the CHOSEN model had a lower AIC 1070 compared to the full model.

As systolic had a stronger correlation than diastolic, we decided to fit two other models to compare to the CHOSEN model from model selection. These were models that included both systolic and diastolic and the model with just systolic. The models we fit are given below, with the model name, number of predictors, percent concordance (PC), and AIC:

Note: Model building terminates because the last effect entered is removed by the Wald statistic criterion.

Summary of Stepwise Selection								
	Effect							
Step	Entered	Removed	DF	Number In	Score Chi-Square	Wald Chi-Square	Pr > ChiSq	Variable Label
1	Sex		1	1	26.9829		<.0001	Sex
2	MCH		1	2	15.4781		<.0001	MCH
3	glucose		1	3	17.1300		<.0001	glucose
4	cholesterol		1	4	8.9480		0.0028	cholesterol
5	LDH		1	5	6.2870		0.0122	LDH
6	diastolic		1	6	6.0317		0.0141	diastolic
7	potassium		1	7	5.4983		0.0190	potassium
8	systolic		1	8	1.3370		0.2476	systolic
9		systolic	1	7		1.1819	0.2770	systolic

(a) Summary

Analysis of Maximum Likelihood Estimates						
Parameter		DF	Estimate	Standard Error	Wald Chi-Square	Pr > ChiSq
Intercept	3	1	-15.9454	2.3753	45.0638	<.0001
Intercept	2	1	-14.6314	2.3600	38.4356	<.0001
Intercept	1	1	-13.6080	2.3486	33.5715	<.0001
glucose		1	0.0264	0.00711	13.7976	0.0002
Sex		1	-1.1859	0.2616	20.5559	<.0001
cholesterol		1	0.00760	0.00254	8.9451	0.0028
potassium		1	0.0644	0.0473	1.8522	0.1735
diastolic		1	0.0234	0.00981	5.6926	0.0170
LDH		1	-0.00993	0.00356	7.7571	0.0054
MCH		1	0.2645	0.0634	17.3969	<.0001

(b) MLE

Odds Ratio Estimates			
Effect	Point Estimate	95% Wald Confidence Limits	
glucose	1.027	1.013	1.041
Sex	0.305	0.183	0.510
cholesterol	1.008	1.003	1.013
potassium	1.067	0.972	1.170
diastolic	1.024	1.004	1.044
LDH	0.990	0.983	0.997
MCH	1.303	1.151	1.475

(c) OR

Association of Predicted Probabilities and Observed Responses			
Percent Concordant	69.8	Somers' D	0.402
Percent Discordant	29.7	Gamma	0.404
Percent Tied	0.5	Tau-a	0.190
Pairs	93033	c	0.701

(d) Association statistics

FIGURE 9.8
Stepwise selection output for final model and model summary

The associated SAS code for fitting these models are:

```
/*CHOSEN MODEL*/
PROC LOGISTIC DATA=heart DESCENDING;
MODEL risk=glucose sex BMI cholesterol cholratio
RBC WBC potassium diastolic  LDH MCH ;
run;
/* MODEL 1 Sytolic and Diastolic*/
PROC LOGISTIC DATA=heart DESCENDING;
MODEL risk=glucose sex  cholesterol
potassium diastolic systolic LDH MCH ;
run;
/* MODEL 2: Systolic only*/
PROC LOGISTIC DATA=heart DESCENDING;
MODEL risk=glucose sex BMI cholesterol cholratio
RBC WBC potassium  systolic LDH MCH ;
run;
```

The output for the chosen model, model 1, and model 2 are below:

CHOSEN model: 7 predictors (with diastolic)

Model Convergence Status
Convergence criterion (GCONV=1E-8) satisfied.

Score Test for the Proportional Odds Assumption		
Chi-Square	DF	Pr > ChiSq
29.6334	14	0.0086

Model Fit Statistics		
Criterion	Intercept Only	Intercept and Covariates
AIC	1150.505	1077.028
SC	1163.847	1121.501
-2 Log L	1144.505	1057.028

R-Square	0.1295	Max-rescaled R-Square	0.1547

Testing Global Null Hypothesis: BETA=0			
Test	Chi-Square	DF	Pr > ChiSq
Likelihood Ratio	87.4766	7	<.0001
Score	76.7383	7	<.0001
Wald	75.9479	7	<.0001

(a) Model convergence and fit

FIGURE 9.9
Chosen model output

Analysis of Maximum Likelihood Estimates							
Parameter		DF	Estimate	Standard Error	Wald Chi-Square	Pr > ChiSq	Partial Correlation
Intercept	3	1	-15.9660	2.3705	45.3659	<.0001	
Intercept	2	1	-14.6529	2.3552	38.7057	<.0001	
Intercept	1	1	-13.6241	2.3438	33.7888	<.0001	
glucose		1	0.0266	0.00710	14.0365	0.0002	0.1026
Sex		1	-1.1552	0.2578	20.0809	<.0001	-0.1257
cholesterol		1	0.00757	0.00254	8.9219	0.0028	0.0778
potassium		1	0.0644	0.0473	1.8555	0.1731	0
diastolic		1	0.0228	0.00979	5.4351	0.0197	0.0548
LDH		1	-0.00979	0.00354	7.6527	0.0057	-0.0703
MCH		1	0.2656	0.0633	17.6256	<.0001	0.1168

(b) Analysis of MLE

Odds Ratio Estimates and Wald Confidence Intervals				
Effect	Unit	Estimate	95% Confidence Limits	
glucose	1.0000	1.027	1.013	1.041
Sex	1.0000	0.315	0.190	0.522
cholesterol	1.0000	1.008	1.003	1.013
potassium	1.0000	1.067	0.972	1.170
diastolic	1.0000	1.023	1.004	1.043
LDH	1.0000	0.990	0.983	0.997
MCH	1.0000	1.304	1.152	1.476

(c) Odds ratio estimates and Wald confidence intervals

Association of Predicted Probabilities and Observed Responses			
Percent Concordant	69.6	Somers' D	0.397
Percent Discordant	29.9	Gamma	0.399
Percent Tied	0.5	Tau-a	0.188
Pairs	93941	c	0.698

(d) Association Statistics

FIGURE 9.9
Chosen model output. *(Continued)*

Model 1: 8 predictors (systolic and diastolic)

Model Convergence Status
Convergence criterion (GCONV=1E-8) satisfied.

Score Test for the Proportional Odds Assumption		
Chi-Square	DF	Pr > ChiSq
30.1877	16	0.0171

Model Fit Statistics		
Criterion	Intercept Only	Intercept and Covariates
AIC	1150.505	1077.944
SC	1163.847	1126.865
-2 Log L	1144.505	1055.944

Testing Global Null Hypothesis: BETA=0			
Test	Chi-Square	DF	Pr > ChiSq
Likelihood Ratio	88.5605	8	<.0001
Score	77.6781	8	<.0001
Wald	76.6315	8	<.0001

(a) Model information

Analysis of Maximum Likelihood Estimates						
Parameter		DF	Estimate	Standard Error	Wald Chi-Square	Pr > ChiSq
Intercept	3	1	-16.0439	2.3723	45.7402	<.0001
Intercept	2	1	-14.7278	2.3568	39.0497	<.0001
Intercept	1	1	-13.6967	2.3453	34.1074	<.0001
glucose		1	0.0252	0.00723	12.1381	0.0005
Sex		1	-1.1710	0.2584	20.5298	<.0001
cholesterol		1	0.00739	0.00254	8.4517	0.0036
potassium		1	0.0662	0.0473	1.9561	0.1619
diastolic		1	0.0152	0.0123	1.5106	0.2190
systolic		1	0.00785	0.00775	1.0256	0.3112
LDH		1	-0.00991	0.00355	7.8073	0.0052
MCH		1	0.2616	0.0633	17.0594	<.0001

(b) MLE estimates

FIGURE 9.10
Model 1 output

Odds Ratio Estimates			
Effect	Point Estimate	95% Wald Confidence Limits	
glucose	1.026	1.011	1.040
Sex	0.310	0.187	0.515
cholesterol	1.007	1.002	1.012
potassium	1.068	0.974	1.172
diastolic	1.015	0.991	1.040
systolic	1.008	0.993	1.023
LDH	0.990	0.983	0.997
MCH	1.299	1.147	1.471

(c) OR estimates

Association of Predicted Probabilities and Observed Responses			
Percent Concordant	69.6	Somers' D	0.397
Percent Discordant	29.9	Gamma	0.400
Percent Tied	0.5	Tau-a	0.188
Pairs	93941	c	0.699

(d) Association statistics

FIGURE 9.10
Model 1 output. *(Continued)*

Model 2: 7 predictors (with only systolic)

Model Convergence Status
Convergence criterion (GCONV=1E-8) satisfied.

Score Test for the Proportional Odds Assumption		
Chi-Square	DF	Pr > ChiSq
24.1753	14	0.0436

Model Fit Statistics		
Criterion	Intercept Only	Intercept and Covariates
AIC	1150.505	1077.521
SC	1163.847	1121.994
-2 Log L	1144.505	1057.521

Testing Global Null Hypothesis: BETA=0			
Test	Chi-Square	DF	Pr > ChiSq
Likelihood Ratio	86.9840	7	<.0001
Score	76.2195	7	<.0001
Wald	75.1924	7	<.0001

(a) Model information

Analysis of Maximum Likelihood Estimates						
Parameter		DF	Estimate	Standard Error	Wald Chi-Square	Pr > ChiSq
Intercept	3	1	-15.5449	2.3346	44.3361	<.0001
Intercept	2	1	-14.2289	2.3192	37.6429	<.0001
Intercept	1	1	-13.1999	2.3079	32.7127	<.0001
glucose		1	0.0248	0.00725	11.7076	0.0006
Sex		1	-1.2322	0.2539	23.5591	<.0001
cholesterol		1	0.00752	0.00253	8.8141	0.0030
potassium		1	0.0684	0.0473	2.0889	0.1484
systolic		1	0.0135	0.00611	4.8835	0.0271
LDH		1	-0.00970	0.00354	7.5008	0.0062
MCH		1	0.2616	0.0634	17.0356	<.0001

(b) MLE estimates

FIGURE 9.11
Model 2 output

Odds Ratio Estimates			
Effect	Point Estimate	95% Wald Confidence Limits	
glucose	1.025	1.011	1.040
Sex	0.292	0.177	0.480
cholesterol	1.008	1.003	1.013
potassium	1.071	0.976	1.175
systolic	1.014	1.002	1.026
LDH	0.990	0.983	0.997
MCH	1.299	1.147	1.471

(c) OR estimates

Association of Predicted Probabilities and Observed Responses			
Percent Concordant	69.4	Somers' D	0.393
Percent Discordant	30.1	Gamma	0.395
Percent Tied	0.5	Tau-a	0.186
Pairs	93941	c	0.696

(d) Association statistics

FIGURE 9.11
Model 2 output. *(Continued)*

To compare these models, we use the model fit measures. Here, we provide the percent concordance (PC) and the AIC values.

- FULL model: 12 predictors, PC=70.1%, AIC=1077
- CHOSEN model: 7 predictors (with diastolic): PC=69.8%, AIC=1077
- Model 1: 8 predictors (systolic and diastolic): PC=69.6%, AIC=1077
- Model 2: 7 predictors (with systolic): PC=69.4%, AIC=1078

Looking at these different options for models, we select the CHOSEN model as our FINAL model. This model Percent Concordance is close to the full model's and there is an AIC advantage. However, any of the other models could have been easily selected depending on what the researchers were looking for. Model selection gives us a set of candidate models. The CHOSEN model, is just that "chosen" by the researcher as the one that fits their questions. Model selection is more of an art than a science, and the underlying questions need to be taken into account as we pick the final model. Stepwise selection allows us to see what the software decides as the model that is "best", keeping in mind that the criterion chosen by software to determine best, may not be the correct criterion. Common sense and underlying questions are all factors in how the CHOSEN model is selected. If the researchers felt that ALL predictors should be included, we would use the FULL model. If we felt that both diastolic and systolic needed to be in the model, Model 1 would be picked. Model selection is a process that combines statistical theory with knowledge of the problem and the interest of the scientists. However, software programs like STEPWISE allow us a direction for coming up with our final model, using an inbuilt optimization criterion like p-value. In addition to using the results of model selection, we should keep also consider the results of our exploratory analysis (for example the results from the LOESS curves on the scatter plots) to make sure all these results make sense in the context of the data.

9.5 Diagnostics

For this section, we focus on diagnostics of the CHOSEN model. Unlike R, all options related with the diagnostics of ordinal logistic model is provided as options when the PROC LOGISTIC syntax is run. Thus, all plots, tests, and association statistics are provided with the model output and are seen in Figure 9.9. However, if there is interest in getting the specific results we will discuss, the following SAS code can be used:

```
1 PROC LOGISTIC DATA=heart DESCENDING;
2 MODEL risk=glucose sex cholesterol  potassium
    diastolic  LDH MCH /
3 pcorr clodds=both ctable lackfit rsquare;
4 run;
```

Below is the output for lack of fit from the above SAS code:

Partition for the Hosmer and Lemeshow Test									
Group	Total	Observed risk = 3	Observed risk = 2	Observed risk = 1	Observed risk = 0	Expected risk = 3	Expected risk = 2	Expected risk = 1	Expected risk = 0
1	63	13	9	15	26	9.58	14.6	15.1	23.6
2	63	3	15	13	32	4.91	10.1	14.4	33.6
3	63	5	10	14	34	3.84	8.40	13.1	37.6
4	63	2	7	11	43	3.16	7.17	12.0	40.7
5	63	1	4	12	46	2.57	6.02	10.7	43.7
6	63	2	4	9	48	2.08	5.00	9.39	46.5
7	63	1	4	8	50	1.66	4.09	8.06	49.2
8	63	1	5	6	51	1.23	3.12	6.47	52.2
9	63	1	3	6	53	0.87	2.26	4.90	55.0
10	64	1	0	3	60	0.43	1.15	2.64	59.8

Hosmer and Lemeshow Goodness-of-Fit Test		
Chi-Square	DF	Pr > ChiSq
14.8340	26	0.9803

FIGURE 9.12
Hosmer-Lemeshow test results

Results from Figures 9.9 and 9.12 show:

1. The proportional odds assumption is tested using the Score Test in SAS. For our CHOSEN model the p-value was 0.0086, hence it was not satisfied. Sadly, none of the models we looked at including the FULL model satisfied this assumption for the dataset.
2. The Hosmer and Lemeshow showed that Goodness of Fit assumptions were not violated (p-value 0.93).

Unfortunately, SAS does not give options for VIF for checking multicollinearity and autocorrelation tests when the code for Ordinal Logistic model used. Hence, we trick SAS into giving us these measures, by running the data as a regression model with the ordinal categories as numbers. As all we are looking for is the correlation among the predictors, the structure of the response is not relevant. Similarly, as we are trying to understand the autocorrelation among the observations in the order they were collected, the nature of response is not relevant. Thus, PROC REG will be used to calculate VIF and the test for autocorrelation. The SAS code an output are below:

```
1 PROC REG data=heart;
2 MODEL risk=glucose sex cholesterol potassium
    diastolic LDH MCH/vif dwprob;
3 run;
```

The REG Procedure
Model: MODEL1
Dependent Variable: risk

Durbin-Watson D	1.754
Pr < DW	0.0009
Pr > DW	0.9991
Number of Observations	631
1st Order Autocorrelation	0.123

Note: Pr<DW is the p-value for testing positive autocorrelation, and Pr>DW is the p-value for testing negative autocorrelation.

(a) Autocorrelation tests

FIGURE 9.13
Tests for autocorrelation and multicollinearity

The results from Figure 9.13 show show that all VIFs are close to 1, showing no evidence of multicollinearity. Further, first-order autocorrelation is 0.123. As we have no idea if the order in which we have the data is the order it was collected, this measure is of little relevance and shown for completion. Hence, The results from Figure 9.13 shows us that this model does not violate the assumptions of multicollinearity and autocorrelation.

Parameter Estimates							
Variable	Label	DF	Parameter Estimate	Standard Error	t Value	Pr > |t|	Variance Inflation
Intercept	Intercept	1	-3.86648	0.77124	-5.01	<.0001	0
glucose	glucose	1	0.01303	0.00278	4.69	<.0001	1.06908
Sex	Sex	1	-0.28737	0.07600	-3.78	0.0002	1.08607
cholesterol	cholesterol	1	0.00242	0.00089428	2.70	0.0071	1.07650
potassium	potassium	1	0.03102	0.02092	1.48	0.1386	1.00625
diastolic	diastolic	1	0.00784	0.00355	2.20	0.0278	1.13692
LDH	LDH	1	-0.00313	0.00112	-2.79	0.0055	1.02622
MCH	MCH	1	0.07825	0.02104	3.72	0.0002	1.00686

(b) Multicollinearity tests

FIGURE 9.13
Tests for autocorrelation and multicollinearity. *(Continued)*

The general R code for the analyses done in this chapter is below:

```
# The imported dataset is called DATA
# The column that has the response variable is called
    response_variable
# The columns that have the explanatory variables are
    called explanatory_variablei, where i represents
    the ith variable

# First, select the variables you need from the
    dataset
DATA = DATA %>%
    select(explanatory_variable1, explanatory_
    variable2, ..., explanatory_variablen)


#########Exploratory#############
library(ggplot2)   # loading package for plots
library(dplyr)    # loading package for data
    manipulations and pipes
# Frequency Table
data_tabulated=as.data.frame(table(DATA$response_
    variable))

# Pie Chart (uses the frequency table)
data_tabulated %>% ggplot(aes(x="", y=Freq, fill=Var1
    )) +
  geom_bar(stat="identity", width=1, color="white") +
```

```
19   coord_polar("y", start=0)+
20   guides(fill=guide_legend(title="Name of Response
       Variable"))+
21   theme_void()
22
23 # Barplot
24 DATA%>%
25   ggplot(aes(x=response_variable, fill=response_
       Variable))+
26   geom_bar(stat="count")+
27   xlab("Name of Response Variable")
28
29 # Scatterplot matrix
30 pairs(DATA)
31
32 # Correlation matrix
33 cor(DATA)
34
35 # Summary table for multiple variables
36 library(tidyverse)  ## loading another data
      manipulation package
37 DATA %>% map_df(.f = ~ broom::tidy(summary(.x)), .id
      = "variable")
38
39 # Scatterplot with loess curve
40
41 DATA %>% ggplot(aes(x= explanatory_variable1, y =
      response_variable))+
42   geom_point()+
43   geom_smooth()
44
45 ########## Modelling #################
46
47 library(MASS)  # loading package for ordinal logistic
       regression
48
49 # Fitting model
50 model_fit = polr(ordered(response_variable) ~
      explanatory_variable1 + explanatory_variable1 +
      explanatory_variable1 + ... , data = DATA, Hess =
      TRUE, method = c("logistic"))
51 summary(model_fit)
52
53 # Saving the R output
54 coefficient_table = coef(summary(model_fit))
```

```
55
56 # Calculating p-values
57 pvalues = pnorm(abs(coefficient_table[, "t value"]),
      lower.tail = FALSE) * 2
58
59 # Saving the p-values and the confidence intervals to
      the original R Output
60 coefficient_table = cbind(coefficient_table, "p value
      " = pvalues)
61 coefficient_table
62
63 # Calculating the confidence interval for the odds
      ratio Estimates
64 Confidence_Interval = confint(model_fit)
65 Confidence_Interval
66
67 # To get confidence interval Assuming Normality, use
      this
68 Confidence_interval_Normality = confint.default(model
      _fit)
69
70 # Calculating odds ratio
71 OR = exp(coef(model_fit))
72
73 # Calculating  confidence interval for odds ratio
74 ConfidenceInterval = exp(confint(model_fit))
75
76 # Putting odds ratio and confidence interval in the
      same table
77 cbind(OR, ConfidenceInterval)
78
79 # Stepwise selection using the AIC criteria
80 stepAIC(model_fit)
81
82 ############# Diagnostics ###################
83
84 # Checking the Proportional Odds Assumption
85 library(brant)  # Loading package to test the
      assumption
86 brant::brant(model_fit)
87
88 # Checking  multicollinearity
89 library(regclass)  # loading library to calculate vif
90 car::vif(model_fit)
91
```

```
# autocorrelation
# In R (Chapter 7), we used Cramer's V, Cohen's Kappa
    and Goodman Kruskal Tau to look at
   autocorrelation.

# goodness of fit test/ lack of fit test
library(gofcat) # Loading the package gofcat
hosmerlem(model_fit,tables = TRUE)
```

9.6 Summary

In this chapter, we analyzed the "Heart" data using SAS and provided generic R Code. We have provided the steps of the logical process and our decisions along the way. Data analysis follows some basic steps, but along the way the decisions that one takes can depend upon the analyst. We tried to go through the science of the process—explaining why we did what we did, how we analyzed, what we found, and how **we** interpreted our results. Our results indicate that a reasonable model for predicting risk consisted of 7 out of the twelve predictors. The model made sense in terms of the interpretations of the variables (i.e., using the signs and strengths of the β). All assumptions, but the proportional odds assumption, were satisfied. The percent concordance for the chosen model indicates a model with fairly good predictability. Goodness of Fit by Hosmer and Lemeshow were satisfied. So, all in all a reasonable model.

10

Tying up Loose Ends and Overall Summary

10.1 Introduction

We started this book talking about why a second look at ordinal data in terms of regression models was warranted. We motivated our book with examples from health sciences (Race Horse and Heart datasets). In this chapter, we will attempt to summarize our overall findings in the context of the problems. This book is meant for researchers, statisticians, and scientists who are looking at an ordinal response and are wanting to understand the models and the assumptions that go with it. As a result, we focused on the **MODEL** aspect of each problem. This book is about modeling ordinal data. Hence, our random variable of interest is an ordinal variable (which is categorical, with some numerical properties that cannot be ignored). However, since there is no distribution that is available for the ordinal variable, we use the binary probabilities of $P(Y \leq j)$ or $P(Y > j)$ and use results related to the multinomial distribution as we saw in chapter 5.

Our purpose for including chapters 4 (Likert) and 6 (SEM) was to understand these techniques from a statistical point of view, in terms of random variables and distributions. Sometimes, when using these models the underlying theory gets muddied in the application, especially since software is available for implementation. In general, both these applications focus on the latent variable, with the interest being in trying to estimate it. In both these ideas, multiple questions are used (with a combination of binary, ordinal and numerical responses). Statistically, the Likert method involves summing or averaging the ordinal responses and assuming that the summing or averaging process allows the Central Limit Theorem to be invoked (with some caveats). Then, appropriate GLMs is be used. The SEM model is more involved as it looks at multiple responses coming from potentially multiple latent constructs, with both stochastic and non-stochastic error structures. Statistical modeling involves multivariate regression (multiple Ys) and Confirmatory Factor Analysis (CFA). The selling point of SEM is the fact that the predictors are allowed to be measured with error. However, assumptions like allowing the errors of the predictors and responses to be uncorrelated allow the application of traditional multivariate regression. To reiterate, these models in general assume continuous latent variables. And methods like CFA are used to address the

DOI: 10.1201/9781003020615-10

issues around ordinal observations from survey data by using factor loadings to *create* the latent construct. Hence, these models are relevant in analysis of survey data with potential ordinal responses.

However, in an observational study (non-survey based study) like ours that has one ordinal response and multiple explanatory variables where the question of interest was to understand the relevance of the explanatory variables in the context of the response and to look at potential predictions, the appropriate statistical method is the CDF model, which this book mainly focuses on.

10.2 Findings from Datasets

As we started the book with the Horse data as the motivating example, it is fitting that we should discuss our entire set of findings here, our logic of analyzing the data and the decisions we made. We will follow up with our findings from the Heart study as well.

Results from the Horse Data

We first briefly describe what the dataset was and what the researcher had asked us to look at. As mentioned in chapter 1, EIPH is a measure of bleeding in the lungs after a race and is fairly common in thoroughbred race horses. While not fatal, it could affect the horse's performance and the researcher was interested in seeing if certain factors could be attributed to higher incidence of EIPH. The researcher shared a few explanatory variables as a part of a much larger study. For this book, the predictors we consider are: distance of the race in meters (continuous), time taken in seconds by the horse (continuous), the horse's starting and finishing position (discrete), type of surface for the race (categorical: turf, dirt, all weather), weather (categorical: clear, cloudy, rainy) and whether they were given a preventive medication (categorical: yes, no). This analysis is for illustration only and the actual analysis is a much bigger study.

Exploratory

This is a crucial step, where we take a long look at data and try and find patterns or issues using graphing or numerical summaries. Our findings for the response Y are:

- Of the 5 levels for the response most of the horses had a EIPH of 0, 1, or 2.
- Only 72 and 15 horses out of the 1071 horses were in categories 3 and 4, respectively.

While, this is good news for the horses, as high EIPH values are not desirable, this does raise some statistical issues. Some of the categories are sparse and often, this could lead to non-convergence of the maximum likelihood estimates (MLE) for analysis.

A logical step would have been combining the last two categories or even the last three and deal with 4 or 3 ordinal responses. However, for illustration we kept all 5 responses categories.

In terms of the explanatory variables:

1. Lasix treatment does not seem to reduce Consensus EIPH based on the interpretations of the plots in section 2.2.2 (a) and (b).
2. The type of surface matters for horses with Consensus EIPH 0 and 1 based on interpretations of plot (d) of Figure 2.2 in section 2.2.2.
3. Lower finish positions were most often with horses that had lower Consensus EIPH (i.e., 0, 1) based on associated interpretations of both plots in Figure 2.5 in section 2.2.3.
4. There is not a clear relationship between finish time and Consensus EIPH based on the three plots in section 2.2.5. In section 2.3.5, we see a slight positive relationship between finish position and EIPH, with a correlation around .1

These are all interesting insights for our model (model_fit) in chapter 5. This model (model_fit) was for illustration purposes only (i.e., to show how an explanatory variable of each type could be used). In reality, the model would include all explanatory variables that are relevant to predicting the response.

Analysis

Our model output including the model estimates, p-values and confidence intervals provided in Figures 5.3, and 5.4 indicate that there is a slight positive relationship between finish time and Consensus EIPH. Finish position is a relevant predictor for Consensus EIPH and both dirt and turf surface were associated with higher Consensus EIPH levels than all weather surface. Lasix treatment as we had conjectured was negatively related to EIPH but was not significant at 10% level of significance.

Diagnostics

Once we have the model, we need to look at how well our model performs in terms of **Goodness of Fit** tests. This leads us to diagnostics. Based on the results from the Horse Dataset, we see from Figures 7.1, 7.2, and 7.3 that Hosmer and Lemeshow and Lipsitz seems to support the correctness of the model (p-values .187 and .11) whereas the PR test appears to reject the correctness of fit (p-value .02). These results are often contradictory as they test different aspects of fit. However, in this case we see that all the p-values are on the smaller side, and we would potentially doubt the correctness of the model.

The **Association Tests** for prediction all yield values that are small and close to 0. These also cast doubt on the goodness of the model, as the predictive property of this model is obviously not good. But if we look carefully at the data we realize that the category 4 has such few responses, it will be hard to get a good estimate of probability in that category. This is potentially the reason the Goodness of Fit and Association tests from chapter 7 show that the model has issues.

We do want to remind the readers that the model was for illustration only and we picked one predictor of each type instead of doing any **model selection** type methods. The association methods in section 7.4 can be used for model selection. We showed how to find AIC and BIC, though by itself these numbers do not give much information. These are helpful when comparing different models that are fit. The best model would be the model that has the lowest AIC, BIC and biggest -2Log L. Though all three metrics may not always agree in terms of the best model.

When we looked at the **proportional odds assumption**, we see that the assumption holds, hence we did not have to use partial proportional odds model for this specific example. However, an example using the partial proportional odds model was shown.

In the Horse Data we do not expect **autocorrelation** to be an issue because the data were not measured over time. This matches up as both measures of autocorrelation used show almost no autocorrelation.

Finally, in looking at **multicollinearity**, we do not see any evidence of issues with the explanatory variables being correlated with each other.

Overall Findings from the Horse Data

Our overall findings suggest that the CDF model used (i.e., model used in chapter 5) is probably not the best model and it has issues around prediction and goodness of fit. There may be other explanatory variables in the dataset that are better predictors of Consensus EIPH.

The purpose was to take the readers through the process, as opposed to finding the best model. Finding the best model for predicting EIPH for race horses is research that is ongoing with this dataset.

Results from Heart Data

Unlike the Horse data that was used in multiple chapters in chunks, we analyzed the Heart Dataset in one chapter. However, for completion we provide a brief overview of our findings.

Exploratory

Exploratory analysis gave us the following observations:

- The response risk is skewed with more people in the low-risk stage. As the sample was from young, healthy individuals this tallies with the background knowledge of the data.
- Not all the twelve explanatory variables showed relationship with the response risk. The ones that showed most relationship were glucose, sex, cholesterol, MCH, LDH, diastolic, and systolic. However, diastolic and systolic also showed relationship among themselves.

Analysis

We performed the analysis in two steps, first finding the "best model" through stepwise selection. Then looking at the CHOSEN model to make sure the directions and strength of the relationships made sense. From model selection we saw that:

- The stepwise option picked the best model with: glucose, sex, cholesterol, MCH, diastolic, and potassium. Systolic that had showed relationship with risk in the correlation plot was not included.
- IF we forced systolic in the model, then neither systolic nor diastolic came up relevant in the model, further emphasizing that those two variables are correlated.
- IF we just keep systolic in the model, and exclude Diastolic we see that the p-values for the predictors Potassium and LDH change as well.

This points clearly to the fact that model selection is an art not a science. And it depends upon the researcher as to which model to ultimately choose as the CHOSEN model. In our case we chose the one suggested by stepwise and kept diastolic in the CHOSEN model and not systolic. But if the researcher thought the systolic made more sense, that would be just as good of a CHOSEN model.

Our results from the CHOSEN model show that six of the seven predictors met the 10% criteria (if we believed that to be the correct level). Potassium, though in the model, had a p-value above this threshold. But as it was incorporated in stepwise, we decided to leave it in. All the signs matched up the LOESS plots.

Diagnostics

The CHOSEN model met most of the assumptions. Goodness of Fit and tests of association pointed to an adequate model. The proportional odds assumption was not met (funny though, if we put systolic instead of diastolic this assumption would be almost met!). The data did not show appreciable multicollinearity or autocorrealtion. All in all, we had a fairly good working model.

10.3 Summary: Final Words

We have given a statistical treatment of analyzing ordinal data using a regression framework. We have discussed several other methods for analyzing ordinal data and have tried to dispel several myths around the analysis of ordinal data generally and ordinal regression. Important things we note:

- Likert analysis does **not** mean taking an individual ordinal response and treating it as a number. Instead, it summarizes (usually averages) multiple Likert items (i.e., multiple ordinal responses relating to a single topic or attitude) into a Likert scale which is then treated as a numerical variable for analysis.
- From a statistical point of view, SEM methods are more related to continuous multivariate regression than ordinal regression as the **latent** construct is assumed to be continuous.
- If the latent construct is assumed to be ordinal, CDF methods like the ones we emphasized in this book are used.
- For ordinal regression with one ordinal response variable, the CDF methods we emphasize in this book are statistically appropriate methods.
- Model selection is by no means an exact science. Software provides us with some directions but the decision is up to the analysts and researchers to decide on the CHOSEN model.

The research in terms of CDF models for ordinal data with multiple numerical explanatory variables still has many open problems. Dealing with dependency is probably the most important open problem. Our hope is with this book, we expose the problems in the field and propel more research in this area.

Bibliography

[1] Alan Agresti. *Analysis of Ordinal Categorical Data*, volume 656. John Wiley & Sons, 2010.

[2] Alan Agresti. *An Introduction to Categorical Data Analysis.* John Wiley & Sons, 2018.

[3] Ana M. Aguilera, Manuel Escabias, and Mariano J. Valderrama. Using principal components for estimating logistic regression with high-dimensional multicollinear data. *Computational Statistics & Data Analysis*, 50(8):1905–1924, 2006.

[4] John Aitchison and Samuel D Silvey. The Generalization of Probit Analysis to the Case of Multiple Responses. *Biometrika*, 44(1/2):131–140, 1957.

[5] John R Anderson. Acquisition of cognitive skill. *Psychological review*, 89(4):369, 1982.

[6] Theodore W. Anderson and Herman Rubin. STATISTICAL INFERENCE IN FACTOR ANALYSIS. 1956. https://digitalassets.lib.berkeley.edu/math/ucb/text/math_s3_v5_article-08.pdf

[7] Philippe Bastien, Vincenzo Esposito Vinzi, and Michel Tenenhaus. PLS generalised linear regression. *Computational Statistics & Data Analysis*, 48:17–46, January 2005.

[8] Ralf Bender and Ulrich Grouven. Using Binary Logistic Regression Models for Ordinal Data with Non-proportional Odds. *Journal of clinical epidemiology*, 51(10):809–816, 1998.

[9] Peter M. Bentler. Some contributions to efficient statistics in structural models: Specification and estimation of moment structures. *Psychometrika*, 48(4):493–517, 1983.

[10] Atanu Biswas and Apratim Guha. Time series analysis of categorical data using auto-mutual information. *Journal of Statistical Planning and Inference*, 139(9):3076–3087, September 2009.

[11] C. I. Bliss. The Method of Probits. *Science*, 79(2037):38–39, 1934.

[12] R. Darrell Bock. Multivariate statistical methods in behavioral research. *Scientific Software International*, 1985.

[13] Harry N. Boone, Jr. and Deborah A. Boone. Analyzing Likert Data. *Journal of extension*, 50(2):1–5, 2012.

[14] R. Brant. Assessing Proportionality in the Proportional Odds Model for Ordinal Logistic Regression. *Biometrics*, 1990.

[15] Michael W Browne. Asymptotically distribution-free methods for the analysis of covariance structures. *British journal of mathematical and statistical psychology*, 37(1):62–83, 1984.

[16] Silvia Cagnone and Roberto Ricci. Student Ability Assessment Based on Two IRT Models. *Metodoloski zvezki*, 2(2):209, 2005.

[17] James Carifio and Rocco Perla. Ten Common Misunderstandings, Misconceptions, Persistent Myths and Urban Legends about Likert Scales and Likert Response Formats and their Antidotes. *Journal of Social Sciences*, 2007.

[18] Grace Chan, Paul W. Miller, and MoonJoong Tcha. Happiness In University Education. *International review of economics education*, 4(1):20–45, 2005.

[19] Dennis L Clason and Thomas J Dormody. Analyzing Data Measured by Individual Likert-Type Items. *Journal of agricultural education*, 35(4):4, 1994.

[20] Samuel DeCanio. Democracy, the Market, and the Logic of Social Choice. *American Journal of Political Science*, 58(3):637–652, 2014.

[21] Ben Derrick and Paul White. Comparing Two Samples from an Individual Likert Question. *International Journal of Mathematics and Statistics*, 18:1–13, 2017.

[22] Robert F DeVellis and Carolyn T. Thorpe. Scale development: Theory and Applications. 1991.

[23] Kimberly A Dukes. Cronhach's Alpha. *Encyclopedia of biostatistics*, 2, 2005.

[24] Roger Eckhardt. Stan Ulam, John Von Neumann, and the Monte Carlo Method. *Los Alamos Science*, 1987.

[25] Morten Fagerland and David Hosmer. How to test for goodness of fit in ordinal logistic regression models. *The Stata Journal: Promoting communications on statistics and Stata*, 17:668–686, 2017.

[26] D.J. Finney. *Probit Analysis: A Statistical Treatment of the Sigmoid Response Curve*. Cambridge University Press, 1947.

[27] Michael Friendly and David Meyer. *Discrete Data Analysis with R: Visualization and Modeling Techniques for Categorical and Count Data.* Chapman & Hall/CRC Texts in Statistical Science. CRC Press, 2015.

[28] Francis Galton. *Natural inheritance.* New York; Macmillan and co., fifth edition, 1894.

[29] Carl F. Gauss. *Theoria Motus Corporum Celestium.* Perthes et Besser, Hamburg, 1809. Translated, 1857, as *Theory of Motion of the Heavenly Bodies Moving about the Sun in Conic Sections*, trans. C. H. Davis. Boston, Little, Brown. Reprinted, 1963; New York, Dover.

[30] Leo A. Goodman and William H. Kruskal. Measures of association for cross classifications. *Journal of the American Statistical Association*, 49:732–764, 1954.

[31] Kruskal W. H. Goodman, L. A. Measures of association for cross classifications III: Approximate sampling theory. *Journal of the American Statistical Association*, 58:310–364, 1963.

[32] William H. Greene. ECONOMETRIC ANALYSIS, Eighth Edition. Pearson, 2017.

[33] Addy Hatch. Heart study toys with fame: Research here could change the way heart disease is identified and treated, 2003.

[34] Trynke Hoekstra, Celestina Barbosa-Leiker, Bruce R Wright, and Jos W R Twisk. Effects of longterm developmental patterns of adiposity on levels of c-reactive protein and fibrinogen among North-American men and women: the Spokane heart study. *Obes Facts*, June 2014.

[35] D. W. Hosmer and Wiley S. Lemeshow. *Applied logistic regression.* John Wiley & Sons, Inc., New York, 1989.

[36] David W. Hosmer and Stanley Lemeshow. Goodness of fit tests for the multiple logistic regression model. *Communications in statistics-Theory and Methods*, 9(10):1043–1069, 1980.

[37] W. Gerry Howe. Some contributions to factor analysis. 1955.

[38] LF Jameson Boex. Attributes of effective economics instructors: An analysis of student evaluations. *The Journal of Economic Education*, 31(3):211–227, 2000.

[39] Karl G Jöreskog. New developments in LISREL: Analysis of ordinal variables using polychoric correlations and weighted least squares. *Quality and Quantity*, 24(4):387–404, 1990.

[40] Karl G Jöreskog and Irini Moustaki. Factor analysis of ordinal variables: A comparison of three approaches. *Multivariate Behavioral Research*, 36(3):347–387, 2001.

[41] Karl G Jöreskog and Dag Sörbom. *LISREL 8: User's reference guide.* Scientific Software International, 1996.

[42] K.G. Jöreskog. A general approach to confirmatory maximum likelihood factor analysis. *Psychometrika*, 1969.

[43] K.G. Jöreskog. *A general method for estimating a linear structural equation system.* In Structural equation models in the social sciences. New York: Seminar, 1973.

[44] W. Keesling. *Maximum likelihood approaches to causal flow analysis.* PhD thesis, University of Chicago, 1972.

[45] Maurice G Kendall. A new measure of rank correlation. *Biometrika*, 30(1/2):81–93, 1938.

[46] Kyungmann Kim. A bivariate cumulative probit regression model for ordered categorical data. *Statistics in Medicine*, 14(12):1341–1352, 1995.

[47] David G Kleinbaum, Lawrence L Kupper, Keith E Muller, and Azhar Nizam. Applied regression analysis and other multivariable methods. 1998.

[48] Martin Knott and David J Bartholomew. Latent variable models and factor analysis, 1999.

[49] D. Lawley. Estimation in factor analysis under various initial assumptions. *British Journal of Statistical Psychology*, 11:1 – 12, 1958.

[50] A.M. Legendre. *Nouvelles méthodes pour la détermination des orbites des comètes.* Nineteenth Century Collections Online (NCCO): Science, Technology, and Medicine: 1780-1925. F. Didot, 1805.

[51] Rensis Likert. A technique for the measurement of attitudes. *Archives of psychology*, 1932.

[52] Angel Lopez-Oriona and Jose A.Vilar. *ctsfeatures: Analyzing Categorical Time Series*, 2023.

[53] Metodi Mazhdrakov, Dobriyan Benov, and Nikolai Valkanov. *The Monte Carlo Method. Engineering Applications.* ACMO Academic Press, 2018.

[54] Peter McCullagh. Regression models for ordinal data. *Journal of the Royal Statistical Society: Series B (Methodological)*, 42(2):109–127, 1980.

[55] C. McCulloch and S. R. Searle. Generalized, linear, and mixed models. 2001.

[56] Richard D McKelvey and William Zavoina. A statistical model for the analysis of ordinal level dependent variables. *Journal of mathematical sociology*, 4(1):103–120, 1975.

[57] Nicholas Metropolis and S. Ulam. The Monte Carlo Method. *Journal of the American Statistical Association*, 1949.

[58] D.C. Montgomery, E.A. Peck, and G.G. Vining. *Introduction to Linear Regression Analysis.* Wiley, 2013.

[59] Jillian J. Morrison. *Washington State University ProQuest Dissertations Publishing*, 2019.

[60] Irini Moustaki. A latent variable model for ordinal variables. *Applied psychological measurement*, 24(3):211–223, 2000.

[61] Martin Moustaki and Irini Knott. Generalized latent trait models. *Psychometrika*, 65(3):391–411, 2000.

[62] Bengt Muthén. A general structural equation model with dichotomous, ordered categorical, and continuous latent variable indicators. *Psychometrika*, 49(1):115–132, 1984.

[63] Bengt. O. Muthen. Beyond SEM: General latent variable modelling. *Siometrika*, 2002.

[64] Linda K Muthén and Bengt O Muthén. 1998–2012. Mplus user's guide. *Los Angeles: Muthén & Muthén*, 2012.

[65] L.K. Muthén and B.O. Muthén. *Mplus: Statistical Analysis with Latent Variables; User's Guide; [version 7].* Muthén & Muthén, 2012.

[66] Monte O. Cheney, Lyle Broemeling, Nairanjana Dasgupta, Peijin Xie and C. Harold Mielke Jr. The spokane heart study: weibull regression and coronary artery disease. *Communications in Statistics - Simulation and Computation*, 2000.

[67] E. S. Pearson. *Mathematical Statistics and Data Analysis.* Belmont, CA: Duxbury, 1938.

[68] Karl Pearson. Mathematical contributions to the theory of evolution. III. regression, heredity, and panmixia. *Philosophical Transactions of the Royal Society of London. Series A, Containing Papers of a Mathematical or Physical Character*, 187:253–318, 1896.

[69] Bercedis Peterson and Frank E Harrell Jr. Partial proportional odds models for ordinal response variables. *Journal of the Royal Statistical Society: Series C (Applied Statistics)*, 39(2):205–217, 1990.

[70] Mary Dupuis Sammel and Louise M Ryan. Effects of covariance misspecification in a latent variable model for multiple outcomes. *Statistica Sinica*, pages 1207–1222, 2002.

[71] Stephen Schilling and R Darrell Bock. High-dimensional maximum marginal likelihood item factor analysis by adaptive quadrature. *Psychometrika*, 70(3):533–555, 2005.

[72] EJ Snell. A scaling procedure for ordered categorical data. *Biometrics*, pages 592–607, 1964.

[73] R. H. Somers. A new asymmetric measure of association for ordinal variables. *American Sociological Review*, 27:799–811, 1962.

[74] C. Spearman. The proof and measurement of association between two things. *American Journal of Psychology*, 15:72–101, 1904.

[75] C. Spearman. *The Abilities of Man: Their Nature and Measurement.* Macmillan, 1927.

[76] Anita L Stewart and John E Ware. *Measuring functioning and well-being: the medical outcomes study approach.* duke university Press, 1992.

[77] Diane E. Duffy Thomas J. Santner. *The Statistical Analysis of Discrete Data.* Springer New York, NYk, 1989.

[78] L. L. Thurstone. Attitudes can be measured. *American Journal of Sociology*, 33(4):529–554, 1928.

[79] L. L. Thurstone. A scale for measuring attitude toward the movies. *The Journal of Educational Research*, 22(2):89–94, 1930.

[80] Louis Leon Thurstone. A law of comparative judgment. *Psychological Review*, 34:273–286, 1927.

[81] Louis Leon Thurstone. Theory of attitude measurement. *Psychological Review*, 36:222–241, 1929.

[82] E R Ugba. *gofcat: Goodness-of-fit measures for categorical response models*, 2022.

[83] Wim J van der Linden and Ronald K Hambleton. *Handbook of modern item response theory.* Springer Science & Business Media, 2013.

[84] Christian H. Weiß. Serial dependence of ndarma processes. *Computational Statistics & Data Analysis*, 68:213–238, 2013.

[85] D.E. Wiley. *The identification problem for structural equation models with unmeasured variables*, pages 69–83. New York: Seminar, 1973.

[86] S. Wright. *Correlation and Causation.* 1921.

[87] Sewall Wright. On the Nature of Size Factors. *Genetics*, 3(4):367–374, 07 1918.

[88] Sewall Wright. The method of path coefficients. *Annals of Mathematical Statistics*, 5:161–215, 1934.

Index

A

Adjacent Category Logit Model, 66
Analysis of Covariance (ANCOVA), 53
Analysis of Variance (ANOVA), 48–53
Association Statistics, 112–115, 182
- AIC, 115–117
- BIC, 115–117
- Goodman-Kruskal Gamma, 113–114
- Kendall's Tau-A, 49, 114–115
- Kendall's Tau-B, 49
- Kendall's Tau-C, 49
- log likelihood, 115–117
- R code, 170–173
- SAS code, 169
- Somer's D, 112–113

Assumptions, 105–107, 132–137, 142–144, 168–173, 80
Attitudes, 48
Autocorrelation, 107, 120–125, 182
- Cohen's Kappa, 122
- Goodman and Kruskal's Tau, 122
- R code, 122–123
- SAS code, 169
- serial dependence, 122–125

B

Barplots, 12–17
- R code, 13–14, 171
- SAS code, 14, 149–150

Binary logistic regression, 57, 64, 120
Binary logit model, 55, *see* also Binary logistic regression
Binary response, 55, 106, 153
Binomial distribution, 53, 62,106
Boxplots, 17–18, 20–22, 80–82
- R code, 18, 22
- SAS code, 19, 22

C

Categorical random variable, 52
Categorical variable, 3–6, 11–26, 70–71, 76, 80–84, 88, 109
Categorization of data, 3–4, 11, 52
CDF models, 40–44, 52–85, 105–106,179
- R code, 67, 171–172
- SAS code, 68, 168

Central limit theorem, 47
Chi square measures, 48, 108–111
Coefficient of determination, 125
Coefficients, 60–62, 69–74, 83–85, 98–99, 120
- simulation, 133–135, 140–142

Concordant, 64, 112
Confidence intervals, 64–65, 69–76, 172, 176
Confirmatory Factor Analysis (CFA), 41–42, 93–95, 174
- R code, 95
- SAS code, 95

Constrained partial proportional odds model, 120
Contingency tables, 24–26, 27, 108–109, *see* also cross table
Continuous probability distribution, 55
- logistic distribution, 55–56
- normal, 55

Coronary atherosclerosis, 6, *see* also Heart data
Correlation, 26–30, 120–121, 152–155
Cross Table, 14–16, *see* also contingency tables
Cumulative probabilities, 57–58, 120, 131–145
Cutpoint, 69

D
Dataframe, 82, 136–137, 143
Deviance, 68, 109
Diagnostics, 105–128
Discordant, 64, 112
Discrete random variable, 52
Discrete variable, 3–6, 19–20, 28–29, 72, 77
Dummy variables, 53, 111

E
Exercise Induced Pulmonary Hemorrhage (EIPH), 5–6, 175–178
  exploratory data analysis, 11–30
  Likert model analysis, 48–51
    R code, 49
    SAS code, 50
  CDF model analysis, 67–85
    R code, 67
    SAS code, 68
  diagnostics, 108–128
    R code, 108, 110, 111, 113, 114, 115, 116
    SAS code, 116
Explanatory variable, 5–6, 52–85, 87–88
Exponential family, 53

F
Factor loadings, 89, 99–100, 127, 175
Frequency distribution, 12–13, 26

G
General Linear Model (GLM), 40, 52–53, 90
Generalized Linear Model (GLiM), 40, 54, 106, 127
Goodness of Fit Tests, 108
  Hosmer Lemeshow test, 108–111, 168–169
  Lipsitz test, 111
  Pulkstenis-Robinson test, 109–110

H
Histograms, 12–13
Heart data, 6, *see* also Coronary atherosclerosis
  CDF model analysis, 155–158
    R code, 171
    SAS code, 169
  diagnostics, 168–170
    R code, 172–173
    SAS code, 168
  exploratory data analysis, 149–155
    R code, 170–173
    SAS code, 153, 154, 156, 159
  model selection, 158–168
    R code, 171
    SAS code, 159

I
independent, 40, 63, 131
inference, 5,7, 64
intercepts, 60–61, 68–69, 90, 134
Item Response Theory, 41, 87

K
Kruskal-Wallis Test, 43

L
Latent variable, 6, 40–43, 51–59,67–69, 86–93, 174
  distribution, 131–144
  error, 67
Latent variable models (LVM), 41, 86–89
  Confirmatory Factor Analysis, 93–95

R code, 95
SAS code, 95
multiple linear regression, 90–91
R code, 91
SAS code, 91
multivariate regression, 91–92
R code, 92
SAS code, 92
Path Analysis, 92–93
R code, 93
SAS code, 93
simple linear regression, 89–90
R code, 90
SAS code, 90
Structural Equation Models, 96–98
R code, 98–99
SAS code, 99
Likelihood, 62–64, 115–117
Likert model, 4, 46–51, 184
attitude scale, 39
background and history, 38–39
difference between scale and item, 39, 46–48
example of, 43–44, 47–51
R code, 50
response, 49
SAS code, 51
LISREL, 42, 87
Location-scale properties, 53
Log odds ratio, 57–58, 61–65,73–75, 83–85
Logistic distribution, 40, 55–60, 67, 139–141
Logit, 40–44, 54–70, 83–85, 106–107
adjacent-categories, 66
baseline, 66
continuation-ratio, 66
cumulative, 58–60, 65–67, 118
Logit transform, 55

M

Maximum likelihood estimation, 62–68, 100,115, 118, 158
Log-likelihood, 62, 116
MPlus, 42–44, 87, 100
Model selection, 158–178
R code, 170
SAS code, 159
Monte Carlo simulation, 129–130
Multicollinearity, 125–128, 168–170
R code, 125
SAS code, 169
Multinomial distribution, 40, 63, 174
Multiple linear regression, 90–91
Multivariate regression, 91–92

N

Natural logarithm, 57, 75
Normal distribution, 39–40, 53–56, 64–67, 83–84

O

Observational study, 175
Odds, 58
Odds ratio, 57–50, 61–65, 69–76
Ordered logit model, 40, 57–65, 133–137, 142–143, *see* also ordinal logistic regression
interpretations, 69–73
R code, 170–173
SAS code, 162, 168, 169
simulation, 139–144
Ordered probit model, 40–41, 66–67, 83–85
interpretations, 83–85
R code, 83
SAS code, 85
simulation, 131–139
Ordering of data, 4
Ordinal logistic regression, *see* ordered logit model
Ordinal probit regression, *see* ordered probit model

P

Partial least square regression, 126–128

Parameters, 59–64, 89–95, 118, 130
Parametrization, 55–57
Pie charts, 11–14
Plots or curves, 79–81
Poisson distribution, 53
Predictions, 77–83
Principal component regression, 126–128
  R code, 128–129
Probability, 54–63, 82–83, 100, 106
  simulation, 131–140
Probability density function, 56
Proportional odds assumption, 60–61, 107, 117–120, 128–131, 177
  Brant Wald Test, 117–118
  Partial proportional odds model, 118–120
  R code, 119
  SAS code, 119
p-Values, 72–74

Q
Qualitative research, 48

R
R Software, 7–8
  changing variable types, 30–31
  dplyr (library), 7, 31–32
  ggplot2 (library), 7, 30–31
  importing datasets, 30
  lavaan (library), 90–93, 95, 98–101
  pipes, 30–31
Racehorse data, *see* also Exercise Induced Pulmonary Hemorrhage (EIPH)
Random error, 52, 89, 106
Random variable, 52–53, 61, 89, 106, 134
Randomly, 129–130
Rank-based methods, 38, 43
Regression models, 54–55, 91, 174
Residuals, 94, 105–106

S
SAS software, 8
  CALIS (PROC), 89–99
  CTABLE (PROC), 68, 119, 168
  importing datasets, 32–33
  LOGISTIC (PROC), 68, 74–76, 83, 117–119
  procedures, 8, 32
  REG (PROC), 50, 90, 169
Scatterplots, 19–22
  LOESS smoothing, 19–22
  R code, 20
  SAS code, 20
Slopes, 52, 60–61, 68, 87–92, 133, 140
Statistical significance, 50, 72–76, 164
Stochastic, 52, 88–89, 92, 94, 134, 142,174
Structural equation model (SEM) for ordinal data, 41–42, 86–101
  Confirmatory Factor Analysis (CFA), 42, 87, 89, 93–94, 174
  linear regression analysis (LRA), 41–43, 49–53, 89–93
  Path Analysis (PA), 41–44, 98
Summary statistics, 22–28, 139
Surveys, 37–38, 42, 46, 47
Symmetric distribution, 47, 55, 66, 135, 141

T
T tests, 51
Threshold model, 54–55
Tolerance, 6, 54–60
Tukey lambda class, 56

U
Uniform distribution, 130, 132, 134, 136, 140, 142–143
Urban myth, 46

V
Variance Inflation Factor, 125
    R code, 126
    SAS code, 172
Violin plot, 17–22

W
Weighted least squares, 64, 100

Z
Z score, 83

Printed in the United States
by Baker & Taylor Publisher Services